THE POST-S

A Systematic Geo

THE POST-SOVIET REPUBLICS:

A Systematic Geography

Edited by
Denis J B Shaw

Longman
Scientific &
Technical

Longman Scientific & Technical
Longman Group Limited
Longman House, Burnt Mill, Harlow
Essex CM20 2JE, England
and Associated Companies throughout the world

Copublished in the United States with
John Wiley & Sons, Inc., 605 Third Avenue, New York
NY 10158

First published 1995

British Library Cataloguing in Publication Data
A catalogue entry for this title is available from the
British Library.

ISBN 0–582–30175–0

Library of Congress Cataloging-in-Publication data
The post-Soviet republics : a systematic geography / edited
  by Denis J.B. Shaw.
      p.   cm.
    1. Former Soviet republics—Economic conditions.
    2. Human geography—Former Soviet republics. I. Shaw,
  Denis J.B. HC336.27.P67   1994
330.947—dc20   94–3586
                                                   CIP

Produced through Longman Malaysia

Contents

List of figures

List of tables

List of contributors

Michael J Bradshaw, School of Geography and Centre for Russian and East European Studies, University of Birmingham

R Anthony French, Department of Geography, University College London

Robert N North, Department of Geography, University of British Columbia

Judith Pallot, Christ Church and School of Geography, University of Oxford

Denis J B Shaw, School of Geography and Centre for Russian and East European Studies, University of Birmingham

Graham Smith, Sidney Sussex College and Department of Geography, University of Cambridge

Preface

The origins of this book lie in a previous epoch, that long and comfortable period of Soviet slumber that politicians subsequently dubbed 'the era of stagnation'. In those sleepy days when the USSR was ruled by Leonid Brezhnev and his immediate successors, it seemed to the present group of western geographers a relatively straightforward task to write a textbook on that country. What we wanted to do was write a text which would take a fresh look at the Soviet Union and avoid the traditional approaches which were characteristic of so many regional geographies. We were untroubled by thoughts that our subject might suddenly change out of all recognition and we would no doubt have been amused by the suggestion that the USSR itself would collapse within a few short years. Then came March 1985 and the advent to power of Mikhail Gorbachev. By the late 1980s it had become clear that the changes which he had helped to initiate were beyond even his control, and by the end of 1991 not only was Gorbachev himself out of a job but the country he had ruled had disappeared from the world map. In its place fifteen independent states are now struggling to come to terms with their new situation and with the sweeping economic and social changes brought about by the communist collapse. It is these changes which are the focus of this book.

Given this background, it is hardly surprising that it took very much longer to write this book than originally anticipated. Had we written it sooner, it would have dated very quickly indeed. Even now we are only too aware that much of what we have written can only be regarded as provisional and that by the time the book appears, events will have moved on. Yet so important and sweeping have been the changes of the last few years, not only for the former USSR itself but also for the rest of the world, that it is important to provide students of geography with an up-to-date perspective. We may need to revise this book sooner rather than later, but in the meantime it presents the most recent information we have. At the same time, those perusing these pages will immediately see that we believe the present cannot be understood or explained without reference to the past, and that social change, no matter how sudden or sweeping, can never wipe the historical page entirely clean.

Eventually it may become anachronistic to attempt to write a textbook which deals with the former USSR as an entity. Already, some readers may feel that too much attention is given to what is or was Soviet, or Russian, and too little to the non-Russian republics with their distinctive histories and cultures. We are very conscious of the difficulty of doing full justice to the many varied nations, peoples and regions which were formerly part of the USSR. This has always been the case and continues to be so now, not only because of the training and experience of the contributors to this book but also because for various reasons information about Russia is generally easier to obtain than about the other republics. We have tried to ensure that our viewpoint is not Russocentric, but we are conscious that we may not have succeeded entirely. At the same time, given the very recent break up of the USSR, it seems inevitable that the Soviet heritage will loom large in most if not all the republics for some time to come.

In transliterating Russian, we have used the system approved by the American Board on Geographic Names which tends to be standard in most geographical publications. Place names are an obvious difficulty. In the post-Soviet era, many place names adopted by the Soviets are being changed yet again. Russian cities are often reverting to their historic, pre-Soviet names. In the non-Russian republics, however, cities are not only discarding Soviet names but are often translating their historic names from their Russian or Russified forms into their native languages. We have tried to use the new names wherever possible and particularly for republics and regions. However, there are cases where either the historical context seems to demand the use of an earlier name or where we have used the Russified or English form to avoid confusion. We realise that this compromise is not entirely satisfactory and that we may have offended some sensibilities, but it seemed the best solution in this transitional period.

This book would never have appeared without the help and encouragement of numerous individuals. I wish to thank the past and present staff of Longman Higher Education. My wife Andrea collected many of the materials for maps and helped with the task of editing and proofreading and I thank her especially. Finally, we wish to thank our students of the recent and more distant past for their enthusiastic interest, and especially those of the post-1985 generation for their patience as what they were studying literally changed before their eyes.

Denis J B Shaw
University of Birmingham

Acknowledgements

Acknowledgement is made for permission to reproduce the following copyright material: Paul R Gregory, Robert C Stuart and Harper Collins publishers for Fig. 9.1 (from Gregory P R and Stuart R C 1990 *Soviet Economic Structure and Performance*. Fourth Edition, Harper Collins Publishers, New York p. 322) copyright © 1986 by Paul R Gregory. Reprinted by permission of Harper Collins Publishers Inc.; J P Cole and Butterworth-Heinemann for Fig. 9.5 (from Cole J P 1984 *Geography of the Soviet Union*. Butterworths, London, p. 393).

1

The post-Soviet republics: Environmental and human heritages

Denis J.B. Shaw

In December 1991 the Union of Soviet Socialist Republics, otherwise known as the USSR or Soviet Union, ceased to exist as a unified state. In its place on the world map appeared its fifteen constituent republics, now to become independent states. The origins of this momentous event are probably to be found in various economic, social and political forces which began to affect the USSR from the mid-1970s. But the immediate causes lie in the policies of economic and political reform (often collectively termed *Perestroyka*) which were introduced by the last Soviet leader, Mikhail Gorbachev, after coming to power in 1985. The purpose of this book is to discuss the geography of the development and ultimate demise of the USSR and the implications of that geography for the future of its successor states.

The present chapter will examine the origins of the USSR and its predecessor, the Russian Empire, and suggest how their territorial growth and economic development pattern are important to an understanding of the problems now facing the newly independent republics. The republics are depicted in Fig. 1.1, while Fig. 1.2 shows the economic regions into which the entire USSR was formerly subdivided and to which frequent reference will be made throughout this book.

The geopolitical heritage

For nearly 70 years, the fifteen republics of the former USSR were united together in the highly centralized Soviet state, arguably one of the most centralized states in the modern world. The USSR came officially into existence in 1922 but was essentially established after the October Revolution in 1917 on the ruins of the pre-existing Russian Empire. In fact, apart from a few western territories, the Soviet Union occupied virtually the same space as its predecessor and its capital was Moscow, the ancient capital of Russia. As later chapters of this book will show, despite the change of name and political structure after 1917, the Soviet Union remained in many respects a Russian-dominated state. The fact that today's republics were developed as a unity throughout the Soviet period has profound implications for their current difficulties, as will be seen. Hardly less important is their history before 1917 as part of the Russian Empire. To understand the republics today, therefore, we must know something of the origins and development of the Russian state.

The Russian Empire of the period before 1917 originated with the Muscovite state, a principality which arose in late medieval times in the middle of the mixed forest vegetation zone of what is today European Russia and which centred upon the city of Moscow. This was not, however, the first state to which the term 'Russian' can be applied. Already in the late eighth or early ninth century AD a loosely organized state had come in to being centred upon Kiev, now the capital of Ukraine. This state is known to history as Kievan Rus' and to this day historians argue about how far this Kievan state had a basically Ukrainian character and how far it was essentially Eastern Slavic, a single culture out of which Russian, Ukrainian and Belorussian cultures eventually emerged. Whatever may be the truth of the case, Kievan Rus' was not to endure. It eventually fell prey to disputes between its various territorial princes and then in the thirteenth century to the invading Mongols. Rus' then divided into two parts, its western or Ukrainian and Belorussian part falling into dependence upon the Lithuanian and then Polish state, and its eastern part becoming dependent on the Mongol-Tatars but gradually being united by Moscow.

The detailed story of the rise of Muscovy need

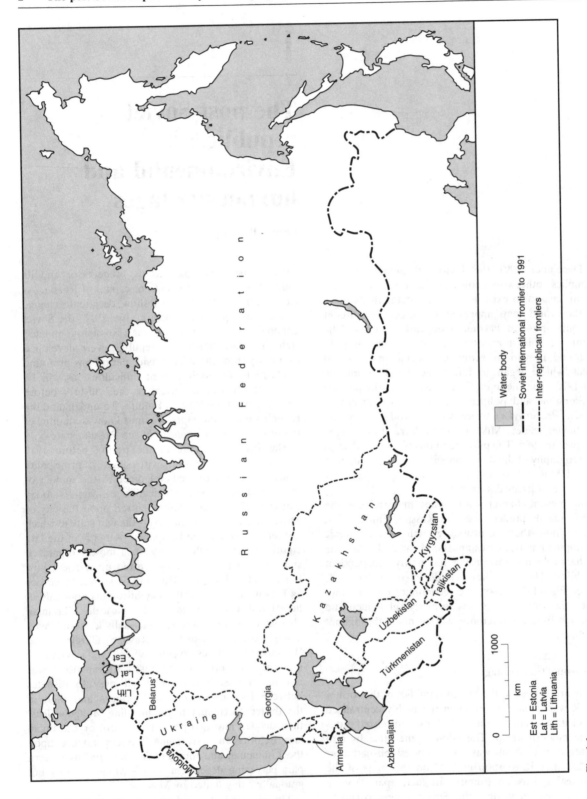

Fig. 1.1
The post-Soviet republics.

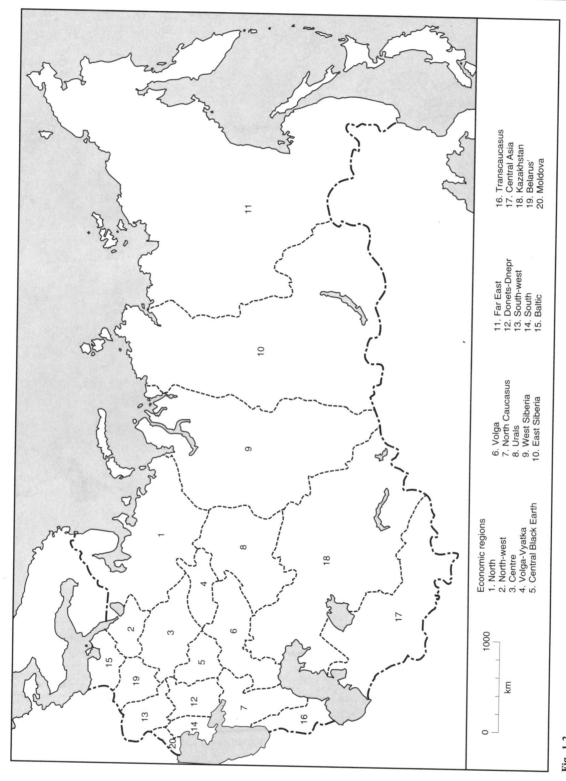

Economic regions

1. North
2. North-west
3. Centre
4. Volga-Vyatka
5. Central Black Earth

6. Volga
7. North Caucasus
8. Urals
9. West Siberia
10. East Siberia

11. Far East
12. Donets-Dnepr
13. South-west
14. South
15. Baltic

16. Transcaucasus
17. Central Asia
18. Kazakhstan
19. Belarus'
20. Moldova

Fig. 1.2
Economic regions of the former USSR.

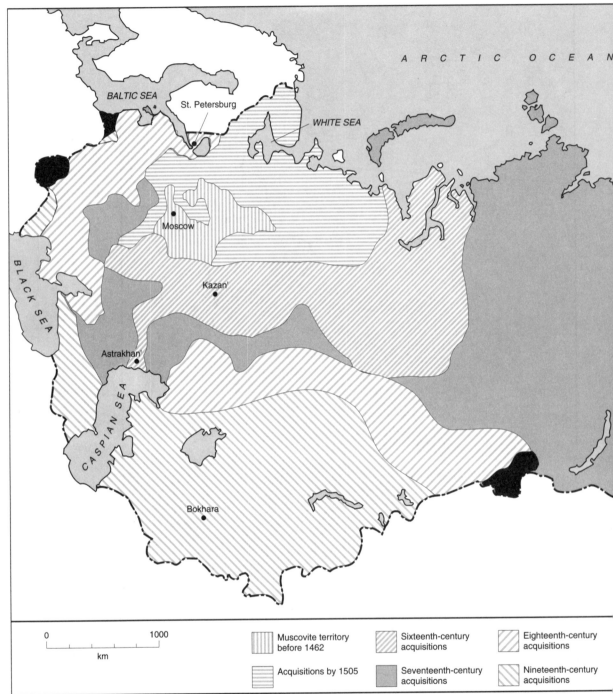

Muscovite territory before 1462	Sixteenth-century acquisitions	Eighteenth-century acquisitions
Acquisitions by 1505	Seventeenth-century acquisitions	Nineteenth-century acquisitions

not concern us. What does concern us, however, is to note the remarkable territorial growth of this state and its gradual transformation into the Russian Empire (Fig. 1.3). According to the Russian historian,

V.O. Klyuchevsky, Muscovy occupied a mere 15 000 square miles in the year 1462, but then expanded at a rate of 50 square miles per day for 400 years (Hunczak 1974: ix). In 1914, the Russian Empire

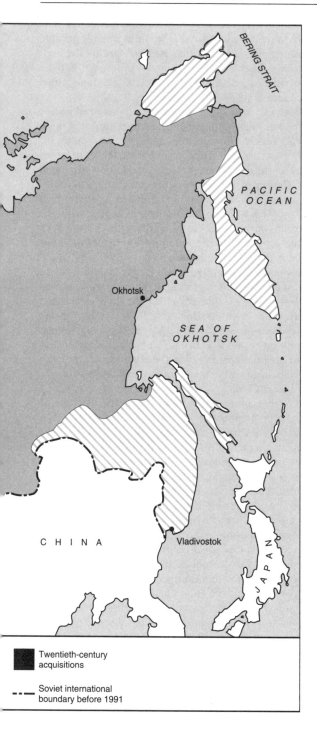

Twentieth-century
acquisitions

- - - Soviet international
boundary before 1991

Fig. 1.3
Territorial growth of the Muscovite and Russian states.

occupied more than eight and a half million square miles, or one-seventh of the land surface of the earth. This expansion occurred in a number of stages. By the latter part of the fifteenth century, Muscovy had annexed most of the other Russian principalities which were mainly located in the mixed forest vegetation zone in the central part of what is now European Russia (see Figs 1.3–1.5). However, it had also fallen heir to the vast regions of coniferous forest to the north, territories which had long been exploited by Russians for their wealth in furs, fish, salt and other resources. In the following century, under the notorious tsar Ivan the Terrible (reigned 1533–84), Moscow conquered the middle and lower Volga valley from the Tatar states of Kazan' and Astrakhan. Russia was now free to begin its vast eastwards expansion across the Urals and the endless coniferous forests of Siberia. This rapid and partly spontaneous movement was motivated above all by the quest for furs and eased by the lack of any sustained opposition from the Siberian native peoples. Already by 1649 the Russians had planted their first settlement on the Pacific at Okhotsk. Then they began to move southwards towards the Amur until checked by the Chinese, and also north-eastwards towards the Bering Strait. In the final maritime phase of the fur trade, the Russians began to operate in the northern Pacific and adjacent parts of North America. Their Alaskan colony was ultimately sold to the United States in 1867.

Whereas Russian expansion to the east was remarkably rapid, that to the west and south was much more difficult. To the west and south-west, for example, lay highly organized states such as Sweden, Poland and Turkey, each one of which was initially at least as strong as Russia itself. To the south and south-east were the open steppe grasslands which were occupied by a number of warlike nomadic peoples. The most important of these nomad groups was the Tatars, who gradually fell into dependence upon their co-religionists, the Turks. Eventually, however, Russia proved powerful enough to challenge all these opponents. The southern frontier of European Russia became a military frontier which gradually moved southwards as Russian military organization eventually proved superior to the raiding skills of the nomadic horsemen. By the late eighteenth century, Russian territory had reached the shores of the Black Sea. In the meantime Russia had also been expanding to the west. Already in the middle of the seventeenth century it had been able to acquire the central and eastern parts of Ukraine from Poland. Many Russians, remembering Kievan Rus', regarded this as a reacquisition of historic Russian territory, but this was a sentiment which was certainly not shared by all Ukrainians. At the beginning of the eighteenth century, under tsar Peter the Great (reigned 1682–1725), Russia acquired lands adjacent to the Baltic Sea in a war against Sweden. It now had secure access to that sea and founded a new port and capital city at St Petersburg. It also secured neighbouring territory which would ultimately become Estonia and the northern part of Latvia.

Later in the same century, especially under Catherine the Great (reigned 1762–96), Russia made large new gains in the west in lands that would eventually become southern Latvia, Lithuania, Belarus' and the western Ukraine. There was no doubt by this stage that the formerly insignificant Muscovite state had transformed itself into a major European power.

Russia's remaining territorial growth occurred mainly in the late eighteenth and nineteenth centuries in three areas. Firstly, in a continuation of the southward movement across the European steppe, Russia annexed territory in the Transcaucasus. Secondly, it expanded into Central Asia across what is now Kazakhstan. This major expansion was at first similiar to that which had occurred across the European steppe, with the Russians having to tackle the warlike nomads who lived in the grasslands to the south of western Siberia, most notably the Kazakhs and the Kyrgyz. By the middle of the nineteenth century, however, the desert lands to the south were being secured by military conquest and Russia was able to annex the long-established oasis states such as Bokhara. In this way Russian territory now approached Afghanistan and China, appearing to menace Britain's position in India. The third area of expansion was the Far East where major new territories were acquired at the expense of a now weakened China.

It is important not to look upon Russia's record of territorial expansion as totally unique. It occurred at exactly the same time that other European states were making territorial gains both in Europe itself and, more particularly, overseas. Thus Russia's gains in the west can be compared with those of its neighbours, Austria and Prussia, as well as with other states in central and western Europe which were competing among themselves for security and space. Elsewhere, Russia's acquisitions can be compared with the vast overseas empires which European powers had been carving out for themselves from the fifteenth century onwards. Unlike other European states, with the partial exception of Austria, Russia had room enough to found its empire in adjacent territories, across the Eurasian landmass. It is also important not to subscribe to myths about Russian expansion, such as to the idea that it demonstrates Russia's uniquely aggressive character. There is no doubt that Russian imperialism had many negative features, and that the colonized peoples often suffered greatly. But all this was paralleled in the imperialism of other countries. Neither can the story of Russian expansion be explained by recourse to any one determining factor, such as a propensity for military aggression or the wish to attain access to the sea. Detailed examination shows that many factors and motives were involved and that the importance of different factors varied both through time and space.

Once again the parallels with other empires are striking (Scammell 1989). Although many attempts have been made to explain European imperialism by reference to such single causes as the search for resources and markets in the expansion of world capitalism, it is very doubtful whether such generalizations can do full justice to the multiplicity of factors involved.

One important result of territorial expansion was greatly to increase the number of people living within the Russian realm. Russia's population has been estimated at about 14 million in 1719, during Peter the Great's reign. By the 1917 revolution, it approached 170 million (Riasanovsky 1969: 307, 478). Equally significant was the fact that it altered the ethnic character of the state. In 1719, for example, the Russians constituted about 70.7 per cent of the population, the Ukrainians 12.9 per cent and the Belorussians 2.4 per cent. By 1917, the Russian proportion was down to about 44.6 per cent, the Ukrainians had 18.1 per cent, the Belorussians 4.0 per cent, and the Poles 6.5 per cent (Rywkin 1988: xv). With the loss of certain western territories in the 1917 revolution, the Russian proportion climbed again to more than half. In 1989, at the time of the last Soviet census, the Russians formed 50.8 per cent of the total.

An important role in the expansion of the empire's population during the eighteenth and nineteenth centuries was played by the natural growth of the ethnic Russian population. Not only did the absolute number of Russians grow greatly but, as was the case in other empires, there was considerable migration and settlement beyond the original homeland. The extent of Russian settlement in newly acquired territories depended upon the natural environment, the degree to which the new lands were settled already and a number of other factors. For example, in the harsh environment of the boreal forests of north, European Russia, comparatively small numbers of Russians settled and intermingled with the pre-existing Finno-Ugrian peoples like the Karelians and the Komi. Much the same picture emerged across the ethnically diverse territories of northern Siberia. Likewise, along the Volga valley to the east and south-east of historic Muscovy Russians settling in the mixed forests of the area mingled with pre-existing Tatar, Mordva, Mari and other peoples.

In the European forest-steppe and steppe, by contrast, the eventual expulsion of the nomads left the fertile lands free for dense Russian agricultural settlement, and here the Russians were joined by Ukrainians, Belorussians and others. Further east, in the more favourable southern parts of Siberia and the Far East, the Russians (as well as other Slavs) also eventually came to outnumber the native populations. In this way Siberia took on a Russian character and much of it became an extension of the historic Russian homeland. In 1911, for example, there were about 8.4 million Russians but fewer than 1 million indigenes in the whole of Siberia and the Far East. One can thus draw a parallel between the Russian settlement of Siberia, and the contemporary white settlement of North America, Australasia and certain other regions.

In many of the more peripheral parts of the Russian Empire, Russian settlement was restricted by the fact that the areas were already well populated by non-Russian peoples. This was true of the Baltic provinces (the future Estonia, Latvia and Lithuania), what was to become Belarus', much of Ukraine apart from the southern steppe, the Transcaucasus, Kazakhstan (except for the north) and Central Asia. However, the overall picture changed in important ways in connection with the tsarist industrialization of the period after 1861 and more particularly in the Soviet era. During these periods of rapid industrialization, the Russians, and to a lesser extent the Ukrainians and Belorussians, proved the most migratory elements in the population. Thus many Russians moved into eastern and southern industrial parts of Ukraine, into northern Kazakhstan, and the Baltic republics (especially after 1945). The Russian element in the populations of the northern parts of European Russia, Siberia and the Far East also grew considerably in connection with the resource-oriented developments of the Soviet period.

In summary, then, Russians were traditionally the majority population in the central European heartland of their expanding empire, and came to outnumber the native populations in some of the other territories, such as much of Siberia, a good deal of the southern steppe, and parts of the north. However, beyond this vast region of predominantly Russian settlement were many territories where non-Russians continued to outnumber Russians, in spite of in-migration by the latter.

Russia controlled its sprawling empire in much the same way that other empires were controlled. Towns and cities, military outposts, networks of roads and eventually a system of railways helped to hold the territory together. The tsars, however, believed more in stick than carrot and paid little heed to the cultural or political aspirations of their varied subjects. Russian cultural norms were upheld, and non-Russians were expected to conform or otherwise to keep quiet. From time to time they were persecuted.

When Lenin and the Bolsheviks (communists) seized control of the state in the revolution of 1917, they were soon faced with resistance as various peripheral regions were occupied by armies opposed to their regime. The Bolsheviks occupied Moscow and the central parts of

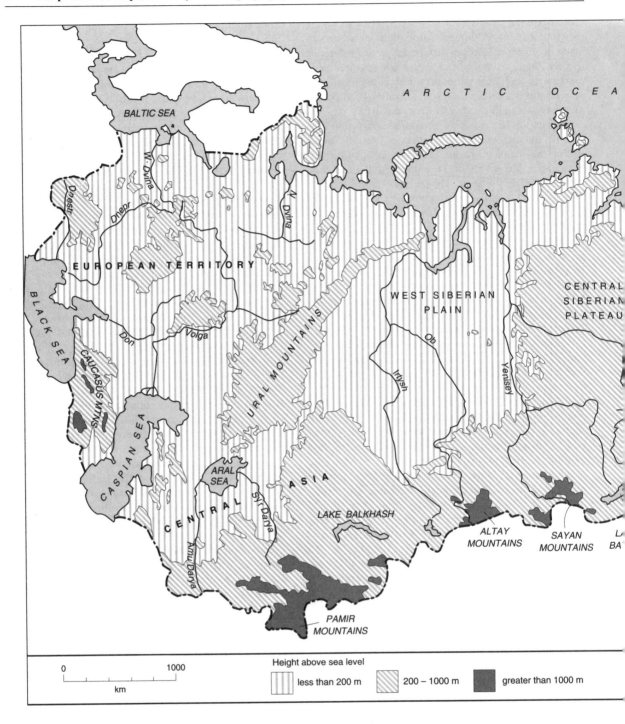

the former empire and it was the infrastructure built by the tsars to hold the empire together (particularly the railways) which helped them to win the ensuing civil war (1918–21). Thus the Bolsheviks eventually found themselves in control of most of the territories formerly ruled by the tsars and it was upon this that they erected their new state. However, unlike the tsars, the Bolsheviks decided to pay some heed to the ethnic

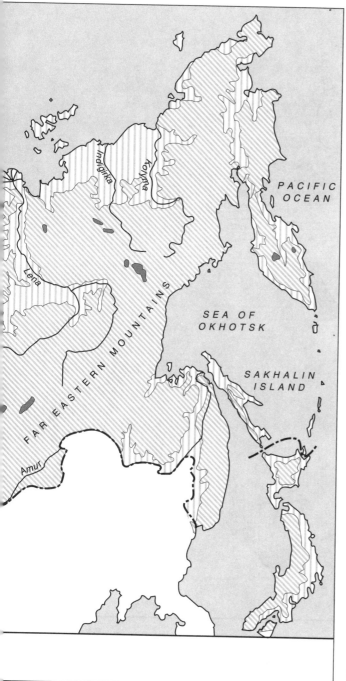

Fig. 1.4
The former USSR: major physical features.

aspirations of their subjects and eventually resolved to design the USSR as a federation of republics and ethnic territories. The implications of this policy will be examined in Chapter 2. What needs to be considered here is the environmental context in which these republics were created, before going on to look at some of the broader factors which were to influence their long-term evolution.

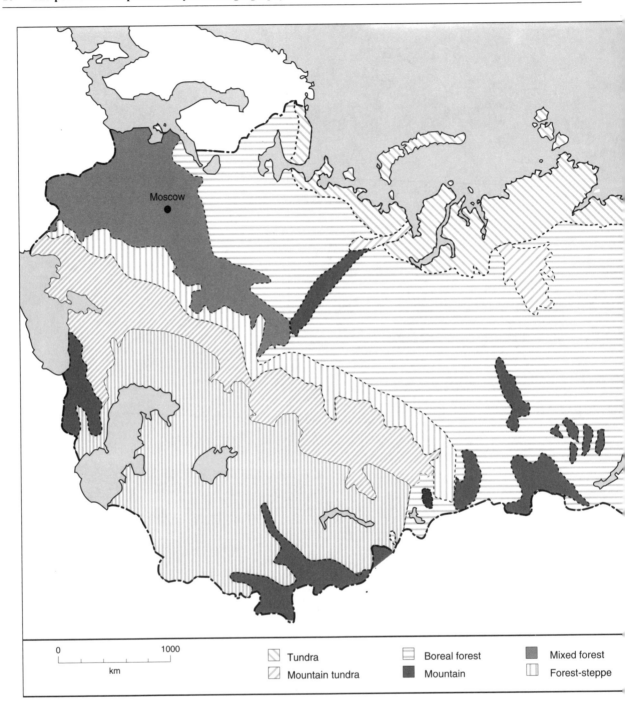

0	1000						

⧄ Tundra   ▤ Boreal forest   ▨ Mixed forest

⧄ Mountain tundra   ■ Mountain   ▦ Forest-steppe

km

Population, environment and resources

Geographers have traditionally been very aware of the relationship between the distribution of human settlement and the natural environment. In a territory as large and diverse as the former Soviet Union, the environmental constraints on settlement become very apparent. This is even evident to some degree from the geographical disposition of the republics themselves (Fig. 1.1). The communists decided to

Steppe

Semi-desert and desert

Far eastern forest

Water body

Fig. 1.5
The former USSR: natural vegetation.

establish republics only for those ethnic groups which were considered sufficiently numerous and territorially compact enough to deserve them. Thus the fourteen non-Russian republics are located on the western and southern peripheries of the former USSR where natural conditions had permitted sufficiently dense human settlement to occur. The territories which were well settled by the Russians themselves formed the core of the fifteenth republic which was created to represent their nationality, known as the Russian Federation.

This then left two huge territories with comparatively sparse populations. The vast and inhospitable northern regions contained a lightly scattered population of Russians and non-Russians and, since none of the latter peoples were sufficiently numerous to be granted their own republic, they were all included in the Russian Federation. The Central Asian deserts were also sparsely populated and it made geographical sense to include them in the republics formed for the non-Russian peoples living mainly in the oasis regions nearby.

Some idea of the environmental constraints on human settlement in former Soviet territory can be gained from a few statistics. Thus only about 10 per cent of the territory can be used for arable farming, with an additional 15 per cent or so for natural pasture. Two-thirds of the territory has less than 10 per cent of the population. About 85 per cent of the population lives on less than a quarter of the land area. The latter zone, containing more than ten persons per square kilometre, forms a triangular region with its base against the former Soviet western frontier and tapering eastwards across the central parts of the former European USSR into West Siberia (see Fig. 6.1). Two significant outliers are to be found in the Caucasian and Transcaucasian regions, and in the Central Asian oases. All these regions correspond with territories having agricultural potential, particularly arable land.

Two factors are of fundamental importance in determining the environmental character of the former USSR: its relatively northerly location within the Northern Hemisphere, and its continental position astride the huge Eurasian landmass. Much of the territory lies far from the moderating influence of the world's oceans and seas. Mountainous terrain, particularly towards the south, south-east and east, shuts out warm air masses which might otherwise come from the Indian Ocean and the Pacific (see Fig. 1.4). The influence of the latter ocean is mainly confined to the coastal areas of the Far East. By contrast, the relatively low relief which characterizes most of the territory means that it is open to air masses coming from the Atlantic or down from the Arctic. The Arctic Ocean is frozen for much of the year, and so this does little to moderate the former USSR's continental character. However, incursions of Arctic air can be important for their negative impact on temperatures. Atlantic influences, on the other hand, are important in these latitudes where westerly circulation patterns tend to predominate. They are the source for much of the precipitation which the territory receives.

The former USSR's continental position is responsible for the climatic extremes which typify so much of the region (Fig. 1.6). Thus winters are cold, generally below freezing for several months of the year. Their severity increases particularly towards the east where the impact of continentality grows; the harsh winters of Siberia require little emphasis here. Oddly enough, however, the degree of discomfort can actually be greater where continentality is less significant. Thus coastal winds and storms in northern Siberia and in the Far East can induce savage wind chills particularly in the winter months. By contrast, summers are warm almost everywhere in the former USSR except in the far north. Average summer temperatures rise towards the south, and especially in Central Asia. But the contrast between summer and winter temperatures is most apparent in East Siberia.

Continentality, especially distance from the Atlantic Ocean, also affects moisture patterns (Fig. 1.7). Most of the territory receives only moderate amounts of precipitation, but there are some exceptions such as coastal areas in the Far East where there is a modified monsoon regime. Generally, rainfall declines as one moves eastwards until one reaches eastern parts of Siberia where the relief produces a more varied pattern. Rainfall also falls away towards the north and towards the south. Moisture deficits are a problem towards the south of former European USSR and in parts of eastern Siberia and the Far East. But they become especially apparent as one moves into Kazakhstan and Central Asia. Here, away from the mountains, semi-desert and desert conditions predominate, exacerbated by the high temperatures of summer. Atmospheric circulation patterns across the former USSR mean that precipitation maxima tend to be in mid-summer in the major agricultural regions, by no means a totally favourable situation for farming. Conversely, lack of snowcover in some of the important farming regions can be a major problem for wintering crops.

Temperature and precipitation together are the major factors that govern the distribution of natural vegetation and hence soils and patterns of human settlement. Figure 1.5 indicates that natural vegetation zones in the former USSR tend to run east–west. The most favourable for agriculture and human settlement are the mixed forest zone, the forest-steppe and steppe, and these largely correspond with the ten persons per square kilometre triangular zone described above. The mixed forest zone lies in the central part of the former European USSR but does not extend east of the Urals. As noted already, this zone together with neighbouring parts of the forest-steppe was the historic settlement region of the Russians. Ukrainians and Belorussians, as well as of the peoples of the Baltic states. Its grey-brown forest soils proved favourable for agricultural settlement. To the south lie the forest-steppe and steppe zones whose rich black earth (*chernozem*) soils, which developed under the former grasslands, were particularly valuable for arable farming once their original nomadic inhabitants had been removed. Their major difficulty is occasional drought. Eastwards the forest-steppe and steppe zones extend across the southern part of West Siberia and the neighbouring northern region of Kazakhstan. However, beyond the Altay Mountains in the south-eastern part of West Siberia, the steppe is confined to a number of intermontain basins. In general, relief is at least as important as climate in limiting the extent of settlement in East Siberia and the Far East.

To the north and south of these favourable zones lie regions which are much less amenable to human endeavour. Immediately to the north lies the boreal forest or tayga whose conifers extend right across north European Russia, Siberia and the Far East. It occupies most of Siberia and the Far East (including the mountains) and altogether accounts for more than half of former Soviet territory. Its poor acidic podzolic soils are rarely useful for arable farming. Further north as well as in mountainous regions of East Siberia and the Far East is the treeless tundra whose natural conditions are even less favourable. Finally, in Kazakhstan the steppe deteriorates southwards into semi-desert and ultimately into the true deserts of Central Asia. The predominantly alkaline soils of this region are also largely infertile, and only in the oases along the major river valleys and along the mountain fringes to the east do conditions permit the existence of intensive agriculture and human settlement.

Naturally, there are many exceptions and additions to this broadly painted environmental portrait. Perhaps the most striking are the so-called 'subtropical' regions of the Transcaucasus where natural conditions are favourable enough to allow the growing of some crops virtually impossible to cultivate in the rest of ex-Soviet territory.

The relationship between the location of the fifteen republics and the environmental patterns described above can now be spelled out (Fig. 1.5). The well-settled and generally agriculturally productive mixed forest zone is shared between the three Baltic republics, Belarus', Moldova, Ukraine and Russia. Ukraine also has a good slice of the even more fertile forest-steppe and steppe. Russia, of course, has far more of the last named types of territory and the steppe also extends into northern Kazakhstan. However, the Russian and Kazakh shares suffer from greater continentality and probably have lower agricultural potentials than areas further west. The boreal forest and tundra lie entirely within the Russian Federation while the deserts are shared between Kazakhstan and the four Central

Asian republics. The inherent importance to both human settlement and agriculture of the oases of southern Kazakhstan and Central Asia and of the 'subtropical' lands of the Transcaucasian republics has been indicated already. Each of the republics, then, has its share of agricultural resources but each faces different environmental challenges and problems in making use of the potential.

The distribution of non-agricultural resources like minerals and energy is obviously closely related to geological factors which cannot be discussed here. Weberian analysis suggests that it is the most convenient resources of this type which tend to be exploited first, and this indeed is what happened in the industrialization of Russia and the USSR. Thus former European USSR, which latterly had about 70 per cent of the Soviet population and 75 per cent of its economic potential, was the initial locus for the exploitation of coal, oil, iron ore and other industrial necessities. Throughout the Soviet period, however, and especially from the 1970s, it became more and more necessary to look further afield as European resources were exhausted. Thus the European north, Siberia, Kazakhstan, Central Asia and the Far East grew increasingly important as resource producers. All this, of course, has taken on a new significance since the Soviet break-up. The western republics find themselves located in regions whose resources are now scarce and expensive because of their long exploitation. By contrast, the northern and eastern territories, falling in the Russian Federation, Kazakhstan and the four Central Asian republics, have the most valuable endowments. The later chapters of this book will show the far-reaching implications of this fact for the evolving political and economic geography of the territory.

The problem of relative backwardness

The heritage which the tsarist and Soviet eras have bequeathed to the fifteen post-Soviet republics is by no means confined to such matters as the character of their natural environments and patterns of settlement. Of equal significance is the structure of their economies, and this in turn relates to their history of development. One of the most important legacies left by both the Russian Empire and the Soviet Union to the successor states is an historic problem of relative backwardness. And like so many other aspects of Russian and Soviet reality, this problem has numerous geographical dimensions.

The rise of the Muscovite state in the late medieval period has been discussed already. One of the many

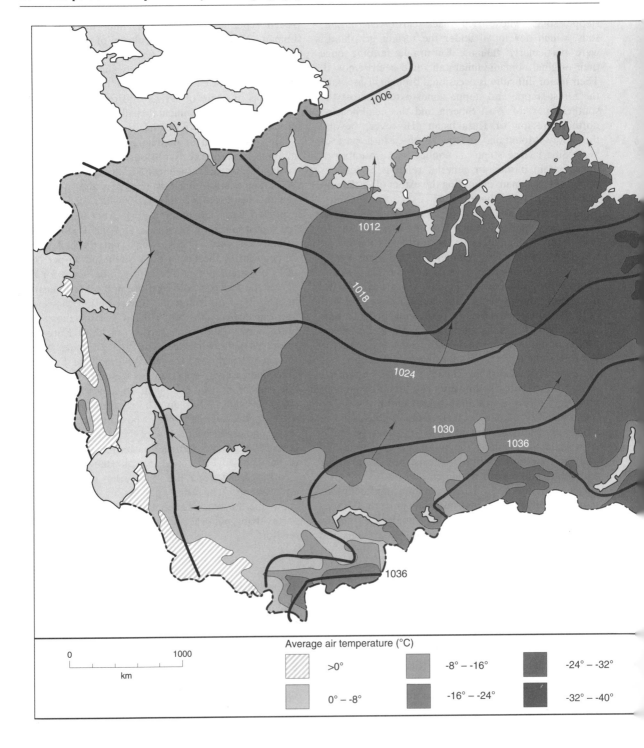

Average air temperature (°C)

>0°	-8° – -16°	-24° – -32°
0° – -8°	-16° – -24°	-32° – -40°

0   1000
km

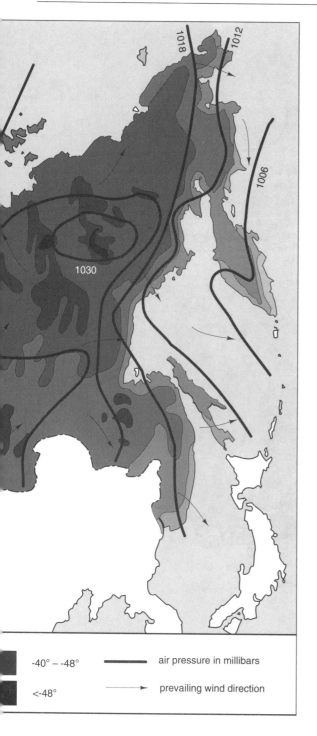

-40° – -48°   ▬▬▬▬ air pressure in millibars

<-48°   ——→ prevailing wind direction

Fig. 1.6a
The former USSR: mean surface air temperatures (°C), sea level pressure and wind directions, January.

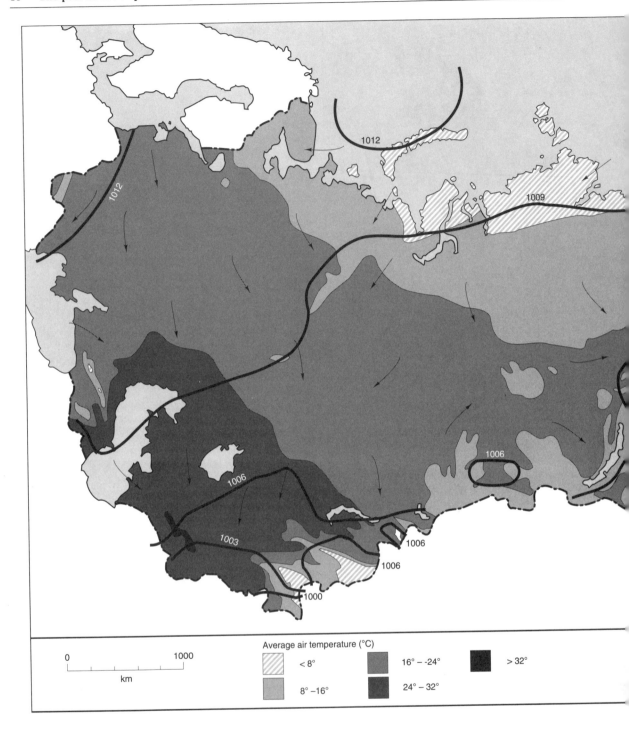

Average air temperature (°C)

< 8°	16° – -24°	> 32°
8° –16°	24° – 32°	

0   1000

km

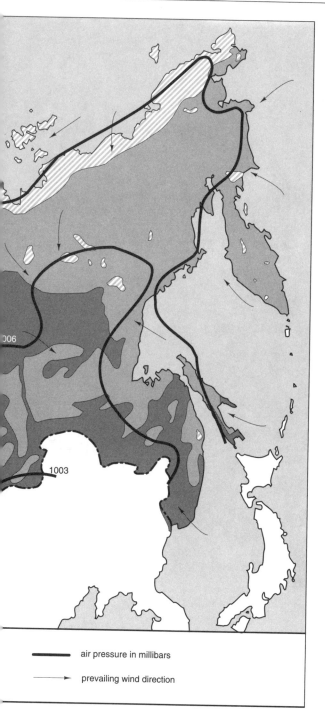

006

1003

——— air pressure in millibars

———▶ prevailing wind direction

Fig. 1.6b
The former USSR: mean surface air temperatures (°C), sea
level pressure and wind directions, July.

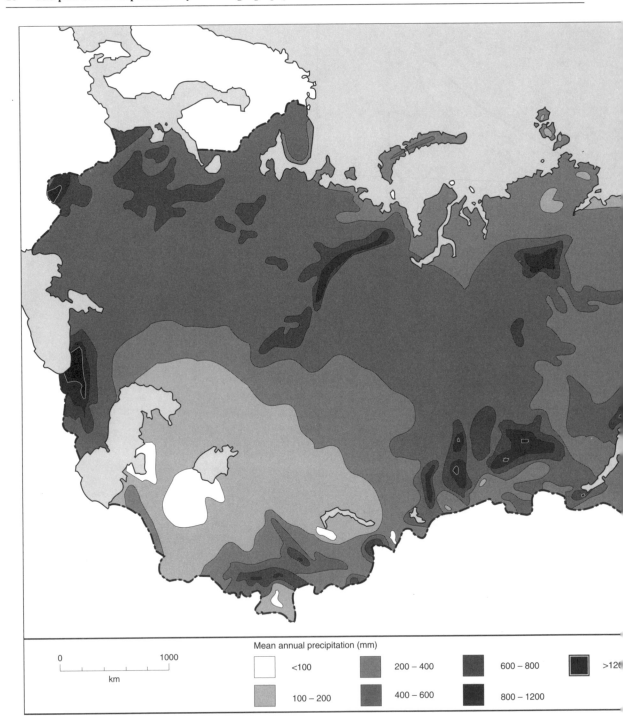

Mean annual precipitation (mm)

<100   200 – 400   600 – 800   >120

100 – 200   400 – 600   800 – 1200

0   1000
km

problems faced by this developing state derived from its location in the centre of the vast plain of European Russia. Open frontiers and the presence of actual or potential enemies on every side made the rulers of Muscovy and Russia acutely aware of the importance of effective defence. For more than two centuries, as we have seen, the princes of Russia were reduced to the status of being mere vassals to the Mongol-Tatars,

Fig. 1.7
The former USSR: mean annual precipitation.

who established an empire known as the Khanate of the Golden Horde in the thirteenth century. Even after the fall of the Golden Horde more than 200 years later, Russia remained vulnerable to the raids of the nomadic Tatars of the steppe. The raids finally stopped only in the eighteenth century.

Russian rulers felt even more vulnerable when faced with threats from the west. In the thirteenth century, at the very moment of the Mongol invasion of Russia, Alexander Nevsky, Grand Prince of Vladimir, had beaten off incursions from the west by the Swedes and the German Order of Knights. Later came contests with the Lithuanians. Poles, Swedes and Turks, all of which were preludes to Russia's emergence as a major European power. Invasions by the French emperor Napoleon in 1812, and by the Germans twice in the twentieth century, were thus the culmination of a whole series of wars and catastrophes in the west.

Growing competition with other European states inevitably made the Russians conscious of the need to match them militarily. Increasing contacts also made them feel that they had much to learn from the West technologically, culturally and in other ways. Disparate factors such as geography itself, religion, and the effects of the Mongol invasion had conspired to distance Russia from the mainstream of European developments, and only from the sixteenth century did Russians become conscious of the disadvantages which this entailed. As early as 1475 the architect Aristotele Fiorovante of Bologna in Italy was employed to build a new cathedral in Moscow dedicated to the Assumption of the Virgin. The cathedral was to serve as a mother church to all Russia and in its Renaissance design and use of constructional techniques new to the country, it was copied throughout the land. In the next century, contacts with the West broadened, especially with the opening up of trading links by way of the White Sea to the north, a route pioneered by the English explorer Richard Chancellor. By the seventeenth and eighteenth centuries, however, Europe's military and technological advantages over Russia were becoming ever more apparent. The Romanov tsars, who ruled after 1613, borrowed ideas more systematically than did their predecessors, but it fell to Peter the Great (reigned 1682–1725) to pursue the first thoroughgoing policy of Westernization. As Paul Dukes writes, with perhaps a modicum of exaggeration: 'In a sense, Russia had not entered modernity before his reign because there was no modernity for it to enter' (Dukes 1991: 8). But this was now to change as Europe began to enter the age of the Industrial Revolution and Russia became one of the first countries to become self-conscious about its relative backwardness. Peter the Great's policies of modernization and industrialization achieved only partial success and failed to provide the basis for sustained, long-term development. Later rulers showed varying degrees of enthusiasm for modernization policies. Two periods of rapid industrial growth require emphasis, however: that between the 1880s and 1917 under the last two tsars, when Russia experienced its own Industrial Revolution; and that which took place under Stalin between the late 1920s

and 1953 which laid the foundations for the industrial development of the later Soviet period.

Despite these major achievements, Russian and Soviet development continued to lag behind that of the West as well as of certain other countries such as Japan. In 1913, for example, national income in Russia amounted to 119 roubles per head compared with 374 in Germany, 580 in Japan, and 1,033 in the USA (Gatrell 1986: 32). Subsequently, Soviet industrial growth reached impressive levels: in 1913, Russia accounted for only about 4 per cent of the world industrial output, compared with about 10 per cent in 1941 and 20 per cent in the mid-1980s (Jeffries 1993: 9). Yet it continued to lag behind the Western powers. In 1987, for example, Soviet GNP was about US$2,375 billion or US$8,363 per capita. Comparative figures for the United States, its principal political rival, were US$4,436 billion and US$18,180 per capita. The Soviet economic base was estimated at only 55 per cent of that of the United States (Gregory and Stuart 1987: 12–14).

Why did Russia and the USSR lag behind in this way? Part of the answer will be found in later chapters of this book. Here we shall look only at some of the long-term factors, a few of which certainly relate to geography. Thus the relatively difficult climate, already discussed at some length, meant that there were always problems for the agricultural population both in feeding itself and in producing the surpluses needed for the towns, for the military, for government and for other purposes. Such problems were exacerbated by the uncertainties associated with disease, famine, warfare and similar catastrophes. In the pre-industrial period, when agriculture was the basis of the economy, such difficulties were a serious hindrance to economic development. There were others. Distance from much of Europe, compounded by long-endured problems in gaining access to the sea, helped shield Russia until late in the tsarist period from the most dynamic influences of a gradually developing world capitalism. The vastness of the territory, problems of ensuring control, and the military burden implied by open frontiers, put an enormous strain on the economy. Resources, while abundant, were often scattered and in difficult and inaccessible locations (Baykov 1954).

Social factors were also important. At least until the late nineteenth century Russia was a rather conservative society. This factor may have been related to the need to avoid risk in a rather unproductive environment (White 1987) as well as to Russia's relative isolation from the outside world. Markets were constrained by poverty and lack of demand, and Russia failed to develop an entrepreneurial middle class on the Western model. For these and other reasons capitalist enterprise before 1917 remained a high risk endeavour. The rigid and hierarchical character of Russian society, though frequently challenged at the local level, was probably sufficient to hold back the development of capitalism. For example, from the 1649 Law Code until the Emancipation in 1861 the mass of the population was subjected to a thoroughgoing serfdom. Enserfment of the populace is probably to be explained by the hard environment and by the difficulties of ensuring control and government revenues in Russian circumstances. Although many social rigidities were being eased by the later part of the tsarist period, others were imposed by the Soviets. It has often been argued that the rigidities associated with the Stalinist planned economy were themselves responsible for holding back development, at least in the long term. This is an issue which is addressed in later chapters of this book.

Given the lag in development, and the apparent need to modernize, it is hardly surprising that some Russian rulers felt the urge to intervene rather than leave their society to its own devices. Russian history provides several spectacular examples of attempts to 'modernize from above', to force an ignorant and generally unwilling populace to move with the times. The first such example is Peter the Great in the early eighteenth century. Peter modernized Russia's armed forces, laid the foundations for the development of many supporting industries, and began the painful process of forcing his reluctant subjects to adjust to European ideas. Peter was a remarkable individual in all kinds of ways, even travelling to western Europe on more than one occasion to learn about Western methods. But like certain of his modernizing successors he imposed much suffering on his people and some of his policies were repudiated after his death. A second example of 'modernization from above' occurred under the last two tsars when the government initially took the lead in developing the railways, attracting foreign capital and protecting domestic industries from international competition. Russia experienced remarkable rates of industrial growth in consequence. Other examples include the Stalin period and, arguably, Mikhail Gorbachev's policy of *Perestroyka* introduced in the late 1980s.

In general, it can be said that Russia's and the USSR's economic development path did not accord with the classic free market model proposed for Britain and some other Western countries. The state played a much bigger role in Russia's case and protectionism was generally important. However, it has been argued by some scholars that pre-1917 Russia's case was not all that unusual (Gerschenkron 1962). Other capitalist countries like Germany and Japan also gave a significant role to the state in economic development and pursued protectionist policies, especially in the initial period of industrialization when they were trying to catch up with more advanced states like Britain. Gerschenkron argued that the Russian state had to play the part of the missing middle classes, raising the capital for investment which in the West was largely done by private entrepreneurs. This interpretation has been challenged by other scholars, but the state certainly felt it necessary to protect Russian industry from foreign competition and to give it other forms of encouragement. It thus largely rejected the free trade doctrines which had strong appeal for countries like Britain which had already established their supremacy in the international market place.

In spite of the success of its industrialization policies before 1917, then, tsarist Russia remained a relatively backward country (see Ch. 4 below). Contemporary commentators worried about whether Russia could nevertheless look forward to future industrial success or whether it was permanently doomed by its peculiar social structure and natural disadvantages to some kind of dependent development (Shanin 1985). Whatever may have been the rights and wrongs of the matter, the contemporary arguments about capitalist development were cut short by the 1917 revolution. After this the country launched itself on an even more distinctive development path than it had been pursuing before. It is to this subject that we must now turn.

Soviet-type development and its present-day consequences

The Bolsheviks who seized power in Russia in 1917 were no less modernizers than their tsarist predecessors. Unlike them, however, they rejected the capitalist development model. Marxism taught that capitalism, based upon private ownership and market relations, was an unjust and exploitative system which confiscated the wealth produced by the majority of the population to the benefit of the minority. Marxists acknowledged the great technological advances which were associated with capitalism and very much believed in the need to continue along the path of industrial progress. But they argued that such progress would be quicker, fairer and more assured if capitalism were to be replaced by socialism. Exactly what that meant in Russian circumstances was uncertain and it fell to the leader of the Bolsheviks, V.I. Lenin, to apply Marxism to Russia. His teachings came to be known as Marxism–Leninism and this was regarded as the official ideology of the Communist Party throughout the Soviet period.

Marxism–Leninism, then, implied that Russia, or the Soviet Union as it was now to become, would continue to industrialize as before but in a completely new, socialist way. At first the Bolsheviks were somewhat uncertain about what that socialist way would be like, especially in view of the fact that the Soviet Union was the first country in the world to have a socialist or communist government. They were also uncertain about what sort of relationship the USSR would have with the rest of the world, much of which was capitalist. The only answer was to feel their way and to respond to circumstances as they arose. In the event, their path was beset with many obstacles and dangers, including civil war, Lenin's death in 1924, and the outbreak of sharp disagreements among the remaining Bolshevik leaders. It was not until the end of the 1920s that the outlines of future policy became clear with the rise to supreme power of Joseph Stalin. Stalin proved determined to abolish Russia's age-old backwardness once and for all. 'One of the features of old Russia', he said in a famous speech in February 1931, 'was the continual beatings she suffered because of her backwardness. . . . In the past we had no fatherland, nor could we have had one. But now that we have overthrown capitalism and power is in our hands, in the hands of the people, we have a fatherland, and we must uphold its independence' (quoted by Fitzpatrick 1982: 118–19). The political and economic system which Stalin created effectively endured until the very end of the Soviet epoch.

The foundations of the Stalinist system were laid by Lenin soon after the revolution with the nationalization of land and heavy industry and the first faltering steps towards the introduction of a planned economy. However, the many uncertainties about policy and the serious blows to the economy produced by the First World War, the revolution and the civil war (1918–21), meant that it was not until the late 1920s that the country's economy had recovered sufficiently to allow Stalin, now firmly in the saddle, to launch his great experiment. What he did was to complete the process of economic centralization initiated by Lenin, effectively abolishing private enterprise in industry and services, collectivizing agriculture and subjecting virtually the entire economy to state planning. All this was accompanied by a strict political centralization with the ruthless suppression of all dissident activity and the indiscriminate use of terror. Henceforward, the accent was to be upon industrialization at all possible speed. The details of Stalin's policies in industry, agriculture and other sectors will be examined in subsequent chapters. Those chapters will also explain how Stalin's successors (notably Nikita Khrushchev, Communist

Party leader from 1953 to 1964 and Leonid Brezhnev, leader from 1964 to 1982) attempted to modify certain of Stalin's economic and political policies, especially the most repressive ones. But it was not until after 1985 under Gorbachev that any serious attempt was made to achieve fundamental change in the Soviet system.

As a result of the 1917 revolution, then, and especially of Stalin's policies, the Soviet Union pursued a distinctive path of economic and social development. That distinctive path produced a distinctive set of spatial patterns and consequently of spatial problems, as subsequent chapters will show. These problems, the heritage of more than 70 years of Soviet-type development, now confront the fifteen successor states. Their success as independent countries very much depends on the degree to which they are able to grapple with the complex issues which they have inherited from the USSR.

One of the most pressing issues which the successor states have inherited from the USSR is the ethnic one, the result of the latter's status as a multi-ethnic polity. This question is examined in the next two chapters.

References

Baykov A 1954 The economic development of Russia. *Economic History Review* Second Series 7(2): 137–49

Dukes P (ed.) 1991 *Russia and Europe.* Collins and Brown, London

Fitzpatrick S 1982 *The Russian Revolution, 1917–32.* Oxford University Press, Oxford

Gatrell P 1986 *The Tsarist Economy, 1850–1917.* Batsford, London

Gerschenkron A 1962 *Economic Backwardness in Historical Perspective.* Harvard University Press, Cambridge, Massachusetts

Gregory P R, Stuart R C 1990 *Soviet Economic Structure and Performance* 4th edn. Harper and Row, New York

Hunczak T (ed.) 1974 *Russian Imperialism from Ivan the Great to the Revolution.* Rutgers University Press, New Brunswick

Jeffries I (ed.) 1993 *Socialist Economies and the Transition to the Market: a guide.* Routledge, London

Riasanovsky N V 1969 *A History of Russia*, 2nd edn. Oxford University Press, Oxford

Rywkin M (ed.) 1988 *Russian colonial expansion to 1917.* Mansell, London

Scammell G V 1989 *The First Imperial Age: European overseas expansion c. 1400–1715.* Unwin Hyman, London

Shanin T 1985 *Russia as a Developing Society.* Macmillan, London

White C 1987 *Russia and America; The roots of economic divergence.* Croom Helm. London

2

Ethnic relations and federalism in the Soviet era

Denis J.B. Shaw

Before the late 1980s, the fact that the Soviet Union was officially a federation of fifteen ethnically distinct republics was often overlooked, especially in the West. The reason for this is quite apparent. From the Stalin period all suggestions of non-Russian nationalism in the USSR had been ruthlessly suppressed and the republics were not much more than names on maps. With the resurgence of nationalism during the Gorbachev period, however, all this changed dramatically. What were previously little more than nominal republics are now independent states. Moreover, within those republics, ethnic minorities of many kinds are currently campaigning for greater rights. In other words, the issues of ethnicity and nationalism, and their relationships to political geography, are now central to an understanding of the geography of the former USSR. This chapter will look at the nationalities issue as it developed during tsarist and Soviet periods. Chapter 3 will examine the more recent picture. First, however, some account must be given of the complex ethnic geography of the former USSR.

The ethnic geography of the former USSR

It follows from what has been said in the previous chapter about the origins of the Russian Empire and the role played in its development by Russian migration and settlement that the Russians constituted a most important element in the ethnic composition of both the Empire and the USSR. In 1989, at the time of the last Soviet census, they formed 50.8 per cent or fractionally over half of the population. Closely related to them are the other Eastern Slavic peoples, the Ukrainians (with 15.5 per cent of the Soviet total) and the Belorussians (3.5 per cent). Altogether then, the Eastern Slavs constituted almost 70 per cent of the Soviet population, a total of 199 million people. A description of the settlement geography of these three

peoples has been given already (see also Fig. 2.1). The point has also been made that all three peoples, but particularly the Russians, played formative roles in the development of both the Russian Empire and the USSR, and have also migrated in considerable numbers right across former Soviet territory.

The remaining 86 million people who formed 30 per cent of the population of the USSR in 1989 consist of a bewildering variety of nationalities and languages. In fact over 120 nationalities were officially recognized. The most important numerically are listed in Table 2.1.

Table 2.1 Major nationalities in the USSR, 1989

Nationality	Number (thousands)	Percentage of total USSR popn.
Russians	145 155	50.8
Ukrainians	44 186	15.5
Uzbeks	16 698	5.8
Belorussians	10 036	3.5
Kazakhs	8 136	2.8
Azeris	6 770	2.4
Tatars	6 649	2.3
Armenians	4 623	1.6
Tajiks	4 215	1.5
Georgians	3 981	1.4
Moldovans	3 352	1.2
Lithuanians	3 067	1.1
Turkmen	2 729	1.0
Kyrgyz	2 529	0.9
Germans	2 039	0.7
Chuvash	1 842	0.6
Latvians	1 459	0.5
Bashkirs	1 449	0.5
Jews	1 378	0.5
Mordva	1 154	0.4
Poles	1 126	0.4
Estonians	1 027	0.4

Source: 1989 Census of Nationality.

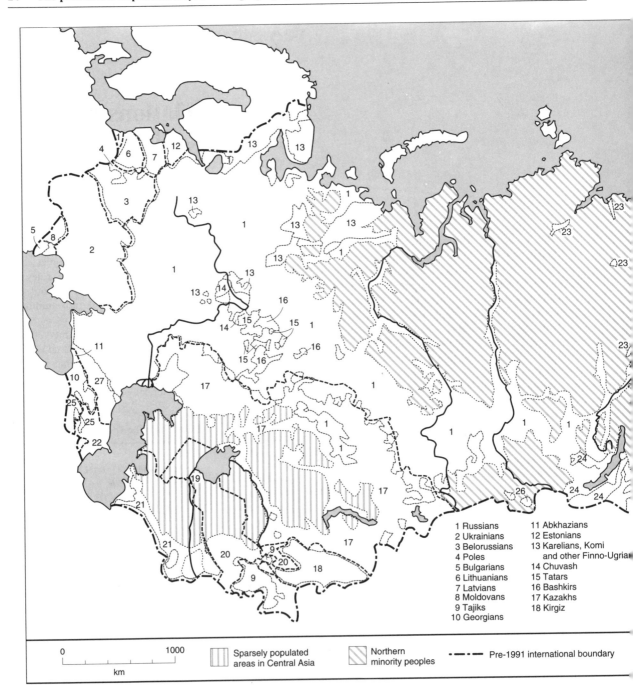

1 Russians
2 Ukrainians
3 Belorussians
4 Poles
5 Bulgarians
6 Lithuanians
7 Latvians
8 Moldovans
9 Tajiks
10 Georgians

11 Abkhazians
12 Estonians
13 Karelians, Komi
   and other Finno-Ugrian
14 Chuvash
15 Tatars
16 Bashkirs
17 Kazakhs
18 Kirgiz

0   1000
km

Sparsely populated
areas in Central Asia

Northern
minority peoples

— · — · — Pre-1991 international boundary

Fortunately, from the point of view of analysis, it is possible to classify nationalities into a number of larger groups.

The Eastern Slavs form part of what is known as the Indo-European group of peoples and languages.

Certain other Slavic nationalities also live in the former USSR, notably the Poles with over 1 million representatives, particularly to be found in Belarus' and Ukraine. Other Indo-Europeans include two of the Baltic peoples (viz. the Lithuanians and Latvians),

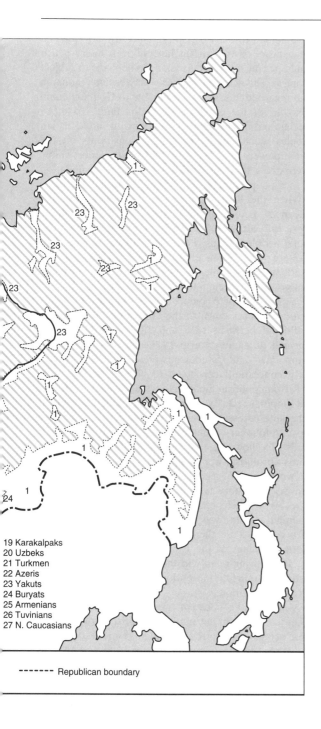

19 Karakalpaks
20 Uzbeks
21 Turkmen
22 Azeris
23 Yakuts
24 Buryats
25 Armenians
26 Tuvinians
27 N. Caucasians

------- Republican boundary

Fig. 2.1
Ethnic geography of the former USSR.

Romanians (notably the Moldovans), Armenians, Jews (who live in many parts of the former USSR but especially in the west), Tajiks (an Iranian group), and Germans. The Germans are descendants of eighteenth-century immigrants who settled mainly in the European

steppe. They were exiled to Siberia and Central Asia by Stalin, but are now moving westwards once more, many of them back to Germany.

Numerically, the second most important group of nationalities is the Altaic group. This in turn can be divided into three. By far the most significant is the Turkic sub-group, who can be found across Central Asia (Uzbeks, Kazakhs, Turkmen, Kirgiz), in the Transcaucasus (Azerbaijanis), in the area stretching from the Volga region and over towards the Urals (Tatars, Chuvash, Bashkirs) and scattered across Siberia (Yakuts, Tuvinians and others). The other two sub-groups are the Mongols (notably the Buryats in East Siberia) and Tungus-Manchurians (especially the Evenki and others in northern Siberia).

Two other important groups are the Caucasians (the Georgians and various other peoples in the North Caucasus and Transcaucasus), and the Finno-Ugrians. The latter are scattered across central and northern parts of European Russia and north-west Siberia. Their most significant representatives numerically are the Estonians, various peoples in the Volga region (Mordva, Udmurt, Mari), the Komi of north-east European Russia, and the Karelians who live near the Finnish frontier.

Minor groups of the former USSR include the Palaeoasiatics (Chukchi, Koryaki) of the north-eastern part of the Russian Far East, the Koreans and the Chinese.

The nationalities in tsarist Russia

The previous chapter indicated that the pre-revolutionary Russian Empire was a direct extension of the Muscovite state. What this implies is that its ruler, the tsar, and the entire political system were imbued with Russian values and cultural norms. Although the tsars were certainly aware of the fact that by no means all of their subjects were Russians, they were prepared to make few concessions to non-Russian ways of life. Indeed it has been argued by some scholars that the expansion of the Russian state across the open plains of European Russia and western Siberia occurred so spontaneously and 'naturally' that at first few Russians were aware of the presence of non-Russian minorities within their realm (Raeff 1971). In other words, few sensed that their country had become an empire. Perhaps this is not entirely surprising in a day when communications were poor and education not widely available. But the particular circumstances attending Russian imperial expansion also affected attitudes towards the non-Russian nationalities. Many of the peoples living north, east and south of the original Russian homeland had more traditional life styles than the Russians themselves and were certainly regarded as culturally inferior. It was therefore understandable that Russians tended to think of the futures of these peoples as involving an ever greater adjustment to Russian ways. There could be no question of somehow giving them freedoms which

were denied to the Russian populace. In any case, no mechanism existed in Russian political theory which would permit special concessions to be made to subject peoples. The Russian tsar was an absolute monarch who regarded himself as answerable to God alone and upholder of the one true Orthodox faith. To make special and permanent concessions to non-Russian peoples would imply permanent constraints on the tsar's powers and suggest that there were viable alternatives to Orthodoxy. An eventual assimilation of such peoples therefore became the ultimate goal.

When the Russians began to rule over culturally 'advanced' peoples along their western borders, however, this attitude became more problematic. The Baltic peoples, Finns, Poles and others were used to Western ideas. Their élites, at least, enjoyed various freedoms and privileges which derived from feudal notions quite foreign to Russian principles. It thus became necessary to compromise to some extent. In the Baltic provinces down to the late nineteenth century, in Poland between 1815 and 1832, and in Finland between its annexation in 1809 and 1899, traditional autonomies and freedoms were upheld. However, these concessions continued to be regarded as alien to Russian political theory and were revoked as soon as the Empire's security appeared to demand it. Although it became necessary to grant further concessions after the revolutionary troubles in 1905, what experience tended to show was that constitutional principles and tsarist absolutism were simply irreconcilable.

Two groups of peoples were subjected to special and discriminatory laws under tsarism since their life styles were felt to contradict the principle of gradual assimilation. One was the nomads. The ultimate goal here was to settle them and thus to bring them into the mainstream of the Empire's life. The other was the Jews, living mainly in the west and having a very distinctive religion and outlook. They were subject to severe persecution.

Although the idea of gradual assimilation was accepted through much of the tsarist period, there were occasions when Russian rulers resorted to policies of active Russification. This became most noticeable in the latter half of the nineteenth century when the Empire's security seemed threatened by various local nationalisms, especially among the peoples in the western part.

The emergence of the USSR as a federal state

'The socialists cannot reach their great aim without fighting against all forms of national oppression' (V.I. Lenin 1915).

The two revolutions of March and November 1917, the first of which overthrew the tsar (to be replaced by a weak Provisional government) and the second of which involved the communist takeover, tore apart the Russian Empire as a unitary state. What eventually emerged to replace it was a federal state, the USSR. This highly eccentric type of federation was the Bolshevik answer to the problem of nationalism.

Nationalism can be defined as the belief that one's people or ethnic group should have its own state and territory (Taylor 1989: 175). While early forms of nationalism are traceable to the late medieval period in European history, its modern form derives from the eighteenth century, especially with the French Revolution and its teaching that political power should rest upon popular consent rather than upon the will of a divinely anointed sovereign. Throughout the nineteenth century, nationalism threatened the stability of many European states, especially large multinational ones like the Russian Empire. Lenin and the Bolsheviks before 1917 were able to make use of nationalist ideas in accusing the tsars of oppressing their national minorities and arguing for the latter's right to 'self-determination'. The Bolshevik belief in national rights was expressed in their 'Declaration of Rights of the Peoples of Russia', issued within a few days of their takeover of power.

However, it must not be inferred from this that Lenin and the Bolsheviks were nationalists at heart. Indeed, they regarded nationalism as an aberration, a purely temporary historical phenomenon which was the product of bourgeois interests and of the many injustices of capitalism. What to them was far more important than nations was social classes, and what they aimed for as socialists was international solidarity between members of the working class. In that sense, then, the Bolsheviks were internationalists who distrusted the divisions which nationalism produced. They looked forward to an international community of working people without any oppression of one ethnic group by another.

Such lofty ideals, however, proved impossible to implement in the realities of Russia in 1917 and succeeding years. The breakdown of authority which followed the fall of tsarism and which continued after the Bolshevik revolution encouraged many nationalities to seize the chance of fleeing Russian domination by declaring their independence. These centrifugal forces were further encouraged by the outbreak of civil war in 1918. Already in the middle of 1917 the Polish bid for independence had been recognized by the Provisional government and was subsequently accepted by the Bolsheviks. Finnish independence followed in December. By the end of 1918, thirteen new states had come into being on the territory of the former empire. Many of them were recognized by the Bolsheviks. In fact the policy of the Bolsheviks was purely pragmatic. They realized that upholding the rights of the nationalities was the best way of ensuring their support in the civil war against their opponents, the 'Whites' and their foreign backers.

In the longer term, however, the Bolsheviks hoped to reassemble the now dispersed peoples into some form of workers' state. In certain cases, notably those of Poland, Finland, and the three Baltic republics of Estonia, Latvia and Lithuania, they were unable to prevent the establishment of independent, bourgeois states. This was in spite of Bolshevik support for workers' movements in those states. Elsewhere, the Bolsheviks were greatly aided by their eventual military victory in the civil war. In many places, and not without Bolshevik assistance, power ultimately fell into the hands of local communists. Because Lenin had managed to organize the Communist Party on a highly centralized basis, he was usually able to ensure Moscow's control even when the communists in question were not Russians. In other places, especially where the communists were weak, it proved necessary to resort to force. In the event the Bolshevik attitude towards the nationalities came to be greatly influenced by military considerations. Lenin was not prepared to see the revolution endangered by the loss of peripheral territories which might then be used by foreign capitalist states (which already supported the 'Whites') to launch strikes against Moscow.

By the early 1920s most of the territory of the former Russian Empire had been recovered by the Bolsheviks with the exception of some of the western regions. On this recovered territory, a series of workers' republics had come into being, controlled to greater or lesser extent by Moscow. Moscow's control was facilitated not only by the centralized structure of the Communist Party but also by other instruments which had been forged and refined during the civil war period, most notably the Red Army, the secret police, and various economic and administrative devices. Nevertheless, there was much uncertainty among the Bolsheviks about how to secure the benefits of unity without at the same time damaging the national sensibilities which were important even to many communists. Stalin, who was Commissar of Nationalities in the Bolshevik government, wanted the other workers' republics to become autonomous republics within Russia (the Russian Federation), the largest of the new workers' states. Lenin, however, feared the consequences of trampling upon national feelings and favoured a federation of nominally equal republics. Eventually, Lenin's view prevailed. In December 1922 a treaty was signed to mark the creation of a federal union between the four republics of Russia, Ukraine, Belorussia (Belarus') and the Transcaucasus. This union became known as the USSR. Needless to say, the amount of autonomy allowed to each of its members was extremely limited.

By the mid-1930s, the number of Union republics had grown to eleven by the splitting up of the Transcaucasian Republic into three (Georgia, Armenia, Azerbaijan) and the addition of five others (Kazakhstan and the four republics of Central Asia). Four more republics (Moldova and the three Baltic states) joined the USSR in 1940. These were situated in territory to the west which had been lost in the civil war but which had now been re-annexed thanks to the notorious non-aggression pact signed with Hitler in 1939. A sixteenth republic, the Karelo-Finnish, was created in 1940 but downgraded in status in 1956 to become an autonomous republic of the Russian Federation.

In the meantime, beginning in 1918–19 and continuing into the mid-1930s, a further federalization was taking place through the establishment of autonomous territories within the Union republics, most notably within the Russian Federation (RSFSR). They included autonomous republics (ASSRs), autonomous oblasts (regions) and autonomous (originally national) okrugs (districts) (Fig. 2.2). These territories were established to grant recognition to the national aspirations of minorities and, with certain changes, they continue to exist to this day. However, there was an important difference in principle between these units and many of the Union republics. The latter were organized on the basis of peoples who had originally seized independence back in 1917–18 and who could therefore claim some kind of equal status with Russia through the 1922 treaty establishing the Union. But the former appeared almost entirely by central decision and were established in such a way as to emphasize their links with the centre. Also part of the policy to recognize the rights of minorities was the establishment across the USSR of a system of national soviets (or councils) at rayon (district) and village level to represent dispersed ethnic groups who lived beyond the bounds of their own republics or autonomous territories, or who had no such territories of their own. These were set up in 1926 but were largely abolished between 1937 and 1939. Their abolition reflected a sharp change of policy towards the nationalities which took place under Stalin.

Soviet federalism in practice

The Soviet Union, therefore, developed as a federal state. But what exactly is federalism and how did it operate in the Soviet case? Political geographers have long been interested in the concept of federalism and have pointed to different types of federal system existing in various parts of the world (Taylor 1989; Paddison 1983; Dikshit 1975). The word 'federal' itself derives from the Latin word *foedus* meaning a treaty, covenant or contract. Inasmuch as the USSR came into being as a result of the treaty signed in 1922, it could be termed a federal state. And yet there are certain problems in applying this concept to the USSR. Many scholars of federalism argue that the term implies a definite and agreed division of powers between two levels of government. However, as noted already, the various instruments of centralization available to the Bolsheviks meant that the amount of autonomy enjoyed by the Union republics was actually extremely limited and uncertain. This fact was already apparent in the 1920s when local autonomist tendencies were quickly pounced on by Moscow. It was to become even more obvious once Stalin had launched his crash programme of economic modernization at the end of that decade. As far as the other national territories were concerned, namely the autonomous republics, oblasts and okrugs, since they were merely created by central fiat it is particularly difficult to apply the concept of federalism to them.

On paper at least the Union republics enjoyed considerable rights. For example, Stalin's constitution of 1936, which was adopted to proclaim the achievement of a socialist economy and political system in the USSR, defined the USSR as a 'federal state formed on the basis of a voluntary association of equal Soviet socialist republics' (Gleason 1992). Each republic was given the right to have its own constitution, to secede freely from the USSR, and to exercise residual powers which had not been assigned to the USSR government. Union republics also enjoyed equal representation in the Soviet of Nationalities, one of two chambers of the Supreme Soviet (the Soviet parliament). Additional provisions adopted in the 1940s stated that republics could create their own military forces and conduct foreign affairs. Two republics (namely Ukraine and Belarus') in addition to the USSR had seats in the United Nations. The republics had many of the trappings of statehood such as flags and institutions of government and numerous other possibilities of expressing their national identities.

Stalin argued that, in order to be able to exercise their rights, all Union republics must have certain characteristics. For example, to be viable, republics must have a population of at least 1 million and their titular nationality must form a majority of that population. Republics must also have a border on the USSR's international frontier in order actually

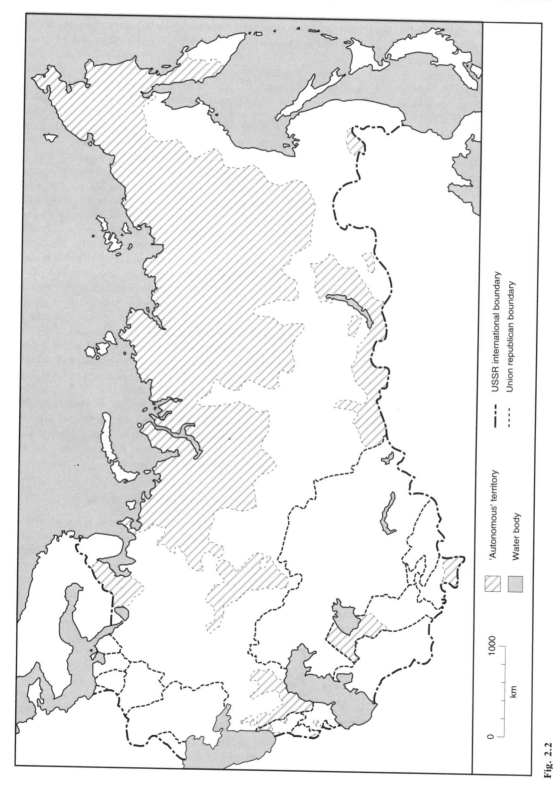

Fig. 2.2
'Autonomous' territories of the former USSR: territory of the ASSRs, autonomous oblasts and autonomous okrugs in the late 1980s.

to be able to secede should they wish to do so. Thus the Tatars on the middle Volga were denied the right to form a Union republic since, in spite of their local dominance and numerical importance, they were surrounded by Russian territory.

In practice, Union republics would have found it almost impossible to exercise their paper rights, since soon after the 1917 revolution, the USSR had become a one-party state with one and the same Communist Party dominant throughout the territory. The party itself was highly centralized, part of Lenin's legacy, and manifestations of dissent within the party were immediately stamped on. Under Stalin, centralization was taken to extremes so that even the exercise of initiative could be a dangerous undertaking. Moreover the party, and by the 1930s Stalin himself, were securely in control of other social institutions. Thus the selection of personnel for important posts, the selection and election of candidates for the Supreme Soviet, and the conduct of elected bodies and state institutions were all subject to strict party control. Education, the mass media and all the means of public persuasion were controlled as were the agencies of coercion such as the police, the secret police and the military. Since almost any form of dissent under Stalin and even later was dangerous, unofficial nationalism found it extremely difficult to secure any place at all in public life.

A further factor which limited republican autonomy was the economic and administrative centralization which occurred particularly from the late 1920s. Stalin's industrialization drive entailed the setting up of a series of central economic ministries in the USSR government, each ministry being responsible for one particular sector of the economy. Throughout the USSR major industrial and other economic enterprises were answerable to their respective ministries in Moscow, irrespective of the republic in which they were located. Although republican governments had responsibilities for regional co-ordination, in practice they had little opportunity to influence industrial activity. Their activities were largely restricted to such matters as consumer services and the local economy. Even here, however, they were constrained by the fact that many services were provided by the economic ministries and their enterprises rather than by republican or local authorities, giving rise to many co-ordination problems at local level. Moreover, even where the republics and institutions of local government had real responsibilities, they were still supervised by the USSR authorities and were forced to operate within controlled budgets.

A slight change to this system occurred between 1957 and 1965 under the influence of Stalin's successor, Nikita Khrushchev. Khrushchev attacked the bureau-cratic ways of the sectoral ministries and replaced them by regional economic councils, controlling industrial activities within each region. He thus moved from what was known as the 'branch principle' of economic administration to the 'territorial principle'. The influence of republican governments was probably enhanced in consequence. However, the reform gave rise to further planning and co-ordination difficulties and was abolished in favour of a return to the 'branch principle' after Khrushchev's fall from office.

Beyond the economic sphere, the republics had a role in socio-cultural matters which also expanded from the mid-1950s. By no means all of this was lost subsequently. In certain aspects of civil and criminal law, in cultural and local planning issues the republics did have room to manoeuvre, always subject to the approval of higher authority. Needless to say, in all matters the freedoms allocated to the lesser autonomous units were even more circumscribed.

The important point about Soviet federalism was that, unlike, say, the American system, there were no real constitutional guarantees which effectively protected the rights of Union republics and of lesser units. The rights which republics and other units exercised were only those which the central authorities allowed, and they could be taken away at any moment. The written constitutions could always be evaded or ignored by the supreme organs of power. Little wonder, then, that some Soviet commentators argued that federalism had lost its purpose by the 1960s since all the Soviet peoples were now merging into one. It is to this issue that we must now turn.

The nationalities and Soviet cultural policy

Lenin regarded federalism as a temporary concession to national sentiment. Once socialism had proven victorious, he believed that nationalism would wither away and that an international proletarian culture would flourish as a result of an international political union of the working class. In the meantime, it was important not to fan the flames of nationalism by proceeding insensitively on national questions. Stalin, as we have seen, was rather less cautious, possibly because he thought more in terms of the USSR's security. Eventually, once it was clear that world revolution would not occur in the near future as many communists had believed, Stalin came to argue that the USSR must build 'socialism in one country'. This was the spirit in which he launched his industrialization drive from the late 1920s, aiming to enhance the industrial and military might of the country. Isolated in the world at large and fearing dissent within, the USSR was beginning by the 1930s to take a much less sanguine view of the nationalities issue than it had done initially.

Lenin's policies and outlook continued to influence the official attitude towards nationalities issues for the first twenty years or so of the USSR's existence. The granting of distinctive territories to the different ethnic groups, a radical departure from tsarist policy (and one which was not approved of by all communists) provided the backdrop to policies designed to foster greater economic and social equality among all the peoples of the USSR. This included policies on economic development, health, education and living standards. All the republics and autonomous regions experienced policies of 'nativization' (*korenizatsiya*), designed to educate local leaders and to recruit them for the great task of building socialism among their compatriots. In this way, indigenous peoples came to play an important role at local level. Encouragement was also given to indigenous languages through education and publishing activity, and numerous languages in remote regions of the USSR were provided with a written form for the very first time. While thus encouraging local cultures, the Bolsheviks, as devout modernizers, had very definite views on what did and what did not count as social progress. For example, religion, traditional medicine and traditional ways of life in general were often treated disparagingly. Thus Bolshevik policies sometimes caused offence, especially in remoter and less developed regions.

Lenin hoped that his 'enlightened' approach to national development among the non-Russian peoples would prove an attractive model for peoples in the non-Western world who might thus be induced to embrace communism. It had the advantage within the USSR of recruiting many non-Russians for the task of building socialism in a situation where there were insufficient educated and trained Russians to send to the periphery. But the policy also had one serious potential disadvantage. Lenin evidently believed that the national development of non-Russian peoples would be a prelude to the ultimate dissolution of national antagonisms. However, by providing the non-Russians with their own territories and making them more aware of their distinctive languages and cultures, he may have been fuelling nationalism in the long term. It could thus be argued that Soviet nationalities policy strengthened feelings of national identity among the non-Russians and in many cases actually created it for the very first time. In other words, it may unwittingly have sown the seeds for the future dismemberment of the USSR.

Lenin's approach to the nationalities issue was gradually abandoned by Stalin. The industrialization drive and the frenetic pace of urbanization were accompanied by mass migration from countryside to town and from region to region. The Russians were the most migratory of the nationalities and many were recruited for the new industries and administrative positions now being created in the peripheral areas, and especially in towns. A considerable amount of ethnic mixing was thus occurring and the principle of *korenizatsiya* diluted or even altogether abandoned. Stalin's economic and administrative centralization undermined local autonomy, as we have seen, and his policies of forced collectivization in the countryside and the denomadization of traditional societies in the far north and parts of Kazakhstan and Central Asia were a severe blow to these cultures. Starting in 1936, Stalin launched a series of purges of officials and party members across the non-Russian regions of the USSR. The victims, many of whom were murdered or sent to prison camps, were accused of various forms of treachery, sabotage, bourgeois nationalism and other crimes against the people. The purged officials were replaced by Russians or by native appointees who were loyal to Stalin, but the effect was further to undermine the principle of *korenizatsiya*. The ultimate departure from Leninist principles came with the forced removal and exile of entire peoples. This began in the 1930s and reached its climax during the Second World War. The major victims included the Soviet Germans, the Crimean Tatars and several North Caucasian peoples. All were accused of collaboration with the Nazis and other false crimes.

About this time Stalin also adopted several policies which amounted in the end to forced Russification. Between 1936 and 1941 most Soviet languages were converted to the Slavic or Cyrillic alphabet used in Russian. The irony is that even languages which had adopted the Latin alphabet in the 1920s, such as those in Central Asia, now had to move to Cyrillic, having thus been forced to undergo two alphabet reforms in less than twenty years. The effect of this policy was to put a greater distance between Soviet languages and related languages outside the USSR (for example, between the Turkic languages of Central Asia and Turkish) and to reduce the distance between these languages and Russian. Thus it now became somewhat easier for non-Russians to learn Russian since the alphabet would already be familiar to them. The adoption of Russian terminology also became more common. Then, by a law of 1938, the learning of Russian became compulsory in all schools. Gradually, therefore, the Russian language and Russian ways were ceasing to be regarded as being on a par with other languages and customs, as they had been during the internationalist 1920s, and were now becoming pre-eminent, much as they were under the tsars. The decline in internationalism reflected the way in which Soviet society turned in upon itself in these years as it began to feel isolated and increasingly vulnerable on the world stage.

The tendency towards Russification took on a new dimension during the Second World War. Before that time, as Stalin consolidated his hold over the country, there had been a growing emphasis on Soviet patriotism. From the late 1930s, however, the specific greatness, historical achievements and importance of the Russian people became a theme. This turned into a major point of propaganda during the war as Stalin, himself a Georgian, struggled to mobilize the Soviet people against the invader. In this way he harked back to an idea which appealed strongly to, if expressed less crudely by, the nineteenth-century tsars.

Stalin had therefore put a whole new gloss upon Soviet nationalities policy. Lenin, as we have seen, was a true internationalist who hoped for an ultimate merger between peoples as equals. Lenin was always extremely critical of Russian nationalism, although he no doubt believed that Russian, as a major language, would have a significant role to play in the future international society. Stalin, however, wanted to see a much more rapid merger between peoples and began to think in terms of assimilation by the Russians of the non-Russians.

This new emphasis was continued by Stalin's successors, although they avoided many of the worst excesses of the Stalin period. Khrushchev, for example, engaged in a certain amount of administrative decentralization to the republics and regions as we have seen. And yet it was also Khrushchev who encouraged talk of a merger (*sliyaniye*) between Soviet peoples and of the appearance of a 'new Soviet person' without specific national affiliation (other than Soviet, despite the fact that after 1932 every Soviet citizen had had his or her ethnic affiliation – Russian, Jewish, Ukrainian etc. – stamped in his or her internal passport). The Party Programme, adopted in 1961, spoke of the borders between Union republics as 'increasingly losing their significance' and of the Soviet peoples as 'united into one family'. It was also under Khrushchev in 1959 that the principle of voluntary choice was introduced into language education. According to this, whereas everyone in the USSR was still required to learn Russian at school, Russians living outside the Russian Federation could now choose to enrol at Russian language schools if they so wished. By this means they could avoid learning the local language. In actual fact, it seems doubtful whether this made much difference, since Russians were notoriously bad at learning local languages in any case (despite regulations, which had been in force since the 1920s, requiring Russian

children living outside the Russian Federation to learn the local language). Its main effect seems to have been a reduction in non-Russian language schooling in Russia itself where, of course, many non-Russians live (Simon 1991: 245–51).

Despite the tendency towards Russification, Khrushchev had permitted a certain flowering of local cultures and this continued after his downfall in 1964. Moreover, official policy in the 1960s and 1970s tended to favour social stability in general terms and, more particularly, stability among local élites. The period thus contrasted with the turmoil and turbulence of previous years. The effect of this policy emphasis was to allow the build-up of local networks of political power which in some regions became notorious for their involvement in the black economy and in mafia-type activity. As long as such élites followed Moscow's line in big issues, they were allowed a fair amount of autonomy locally. Other processes also threatened the party-espoused goal of 'drawing together' (*sblizheniye*) – the term now officially used in preference to the more contentious 'merger' – between Soviet peoples. For example, better education and greater social mobility among native peoples meant that they were now more able to compete with Russians for jobs locally, especially in view of their command of local languages. After 1970 there was a noticeable movement of Russians back to the RSFSR, particularly from several of the southern republics.

It may have been awareness of such processes which spurred the Brezhnev regime to favour Russification measures once more. In debates leading up to the adoption of a new constitution in 1977, for example, there had been talk of doing away with federalism entirely, or of severely reducing the status of republics. In the event there was almost no change, presumably because of the pressure of non-Russian opinion. Also from the late 1970s measures were taken to increase the amount and the quality of Russian taught in schools. The authorities were evidently becoming aware that the desired 'drawing together' between Soviet peoples was taking very much longer than had been hoped. It seems likely, indeed, that nationalism was on the increase once more. The economic inequalities between regions were becoming more obvious as the Soviet economy began to slow down in the 1970s and as the government turned its attention to more pressing matters. Resentment at the local ecological consequences of Soviet development policies was also growing, and Russian immigration into the Baltic states in particular worried many who were concerned about the preservation of their national cultures. The stage was thus set for the explosion of nationalism which occurred in the late 1980s.

Conclusion

In spite of Lenin's hopes for the creation of a new, international community under socialism, it had become clear by the 1970s and 1980s that this would be very difficult to achieve. It could be argued that it was Stalin, in abandoning the Leninist version of internationalism, who had set the stage for the future nationalist upsurge, but the upsurge was also encouraged by other factors which became apparent from the 1970s. By the end of the 1980s, under the impact of Gorbachev's reforms, the Union republics, which had been subservient to Moscow for so long, were demanding a real measure of power. Their example was soon being followed by the other autonomous units, and even by regions and cities having no ethnic basis to support their claims. The entire Soviet system, in other words, was beginning to fragment, and this is a process which has continued into the post-Soviet era. The story of that fragmentation and its consequences is told in Chapter 3.

References

Dikshit R 1975 *The Political Geography of Federalism*. Macmillan, Delhi

Gleason G 1992 The evolution of the Soviet federal system. In Debner R (ed.) *The Soviet Nationality Reader*. Westview, Boulder, pp. 107–20.

Paddison R 1983 *The Fragmented State: The political geography of power*. Blackwell, Oxford

Raeff M 1971 Patterns of Russian imperial policy toward the nationalities. In Allworth E (ed.) *Soviet Nationality Problems*. Columbia University Press, New York, pp. 22–42

Simon G 1991 *Nationalism and Policy Towards the Nationalities in the Soviet Union*. Westview, Boulder

Taylor P J 1989 *Political Geography: World-economy, nation-state and locality*, 2nd edn. Longman, London

3

Ethnic relations in the new states

Graham Smith

The fragmentation of multi-ethnic empires and their replacement by new nation-states is only too rarely accomplished in an orderly and peaceful fashion, particularly when, as in the case of the break-up of the Soviet Union, this process is sudden and unplanned. Rather than satisfying ethnic and national aspirations, for the fifteen post-Soviet states decolonisation has resulted in the proliferation of ethnic conflicts. In part this is linked to their multi-ethnic character in which the nation which legitimized the claim to and laid the foundation for independent statehood has not always been tolerant towards other ethnic groups within the new polity. Consequently, many ethnic minorities fear that their economic, political and cultural well-being is threatened. Fearful of becoming 'victim groups', many have chosen to leave, creating a scale of emigration which in the four Central Asian republics alone may reach 3 million by the end of the 1990s (*Moskovskiye novosti*, 1992). Independent statehood has, however, also had another consequence. In elevating the Union republic borders to the status of international borders, many of which were haphazardly drawn up during the Soviet period, ethnic groups often find themselves excluded from the polity in which the majority of their co-nationals reside. At present there are well over 100 such unresolved inter-state disputes in which the boundaries between ethnic groups do not match territorial divisions.

This chapter examines how national and ethnic relations are shaping the process of decolonisation. First, we explore the transition to nation-statehood and the conditions which gave rise to the formation of these nation-states. Second, we examine how the one-time metropole of the Soviet Empire, Russia, is managing its own multi-ethnic policy through a revitalised form of federalism. And finally, we consider how national and ethnic identities are being reshaped in the non-Russian states and the geopolitical consequences that new state-building practices are having on their sizeable Russian minorities.

The transition to nation statehood

Although the beginning of the end of the Soviet Empire can be located in the late 1980s, the root causes of ethnic discontent in those Union republics which were to emerge as nation states in the autumn of 1991 was systemic, grounded in a particular form of territorial rule which we can label federal colonialism (Smith 1989, 1990a,b). While Moscow had ensured through the maintenance of an ethno-federal structure the preservation and reproduction of ethno-cultural diversity, it also denied those nationality groups which made up the federal structure any meaningful autonomy over their political or economic affairs. So what in effect Moscow had institutionalized was a contradictory set of policies which were both federal and colonial in nature. On the one hand, federalism helped to preserve a sense of ethno-cultural difference through providing the Union republics with a variety of institutional supports for their national cultures. Provided they remained within their own Union republics, natives did not have to assimilate into the Russian language and culture and so abandon their identific links or cultural practices associated with their namesake Union republics. Among other things this ensured the growth of a large native middle class whose identities and concerns were bound up with the preservation of their local cultures and who were to play a central role during the late 1980s in openly questioning the continuation of Soviet rule. On the other hand, denying the Union republics any meaningful say in the running of their own affairs preserved the centralization of party–state power over Union republic affairs. It was this denial of national self-determination among peoples with a well-developed and institutionalized sense of ethno-cultural difference that was to prove the Soviet Union's undoing.

What provided the trigger which was to lead to the eventual emergence of sovereign states was the emergence in March 1985 of a reform-minded leadership under Mikhail Gorbachev. Although beginning as 'a revolution from above' based on far-reaching reforms designed to revitalize the national economy, it quickly became 'a revolution from below' in which through purposely inviting the localities to participate in change, the Union republics quickly took a lead in dictating its nature and tempo. Three such developments were particularly crucial to ensuring their transition from federal colonialism to nation statehood.

First, there was the economic failure of Gorbachev's plans for improving regional living standards. Gorbachev had envisaged an economic reform programme which would have revitalized the economies of the Union republics and which would have arrested falling living standards. *Perestroyka* failed to live up to the high expectations which it had generated. Things in effect got worse. In the non-Russian republics frustration levels rose and were vested with ethnic meaning. It began in late 1986 with riots in Kazakhstan. Although triggered off by Moscow's decision to replace the republic's First Party Secretary with a Russian, discontent within Kazakhstan had much to do with the poor state of the Kazakh economy and the comparatively poorer living standards of Kazakhs compared with Kazakhstan's large (two-fifths) Russian community. By 1990, however, ethnic unrest had not only spread to the other non-Russian republics but had also become more radical and organized in its demands. A number of republics, notably the Baltic republics, Georgia and Azerbaijan began openly to question the merits of remaining part of a Soviet Union in which *Perestroyka* had failed to introduce economic policies designed to improve their living standards or to devolve to their republics sufficient fiscal powers to effect their own economic recoveries.

Secondly, there were the geopolitical consequences of a reform-minded leadership committed to greater democratization. Gorbachev literally invited the Union republics, through introducing *Glasnost'* (greater openness) and pluralist elections to participate in reforming communism. With the lifting of restrictions on the formation of local political associations, grassroots-based ethnoregional movements (or so-called Popular Fronts) began to emerge, first in the Baltic republics in 1988 and later in Transcaucasia, Ukraine and Central Asia. To varying degrees these ethnoregional movements began to call for their peoples' right to greater national self-determination, if not outright independence, for their republics. With the holding of democratic elections in the republics in 1989, the ethnoregions were for the first time in Soviet history also provided with the formal means to articulate decades of pent-up grievances against Moscow-style rule. The republics in effect began to take greater control over their own local affairs, the most successful being those republics, like the Baltic republics, Georgia and Armenia, which quickly marginalized their own local Communist Parties, in the process severing the main political lifeline that the centre had over regulating the pace of change in the republics.

Finally, there was Gorbachev's mishandling of nationalities affairs. It was only as late as 1989 that the Gorbachev administration began to put forward proposals to restructure the Soviet federation, in effect to take on board some of the ethnoregional demands that were coming not only from the non-Russian Union republics but also from a Russian republic resentful of being left behind in the struggle for more say in the running of its own territorial affairs. What in effect Moscow proposed was 'a federation of socialist renewal'. Instead of arguing that federalism was incompatible with state socialism, the Gorbachev administration argued that the original Leninist idea of a federation of equal states was a sound basis upon which to reconstruct the Soviet federation, a principle which, it was contended, had been lost as a result of Stalinist centralization. It was to be a federation based on the formula: 'Without a strong union there are no strong republics; without strong republics, there is no strong union' (*Pravda* 1989). This attempt to accommodate the Union republics was, however, too late. Not only were many of the republics demanding more territorial autonomy than Moscow was prepared to deliver, but a number of republics wanted to go beyond any form of Soviet federalism and establish their own nation states.

It would, however, be wrong to view the non-Russian Union republics as universally committed to establishing their own sovereign states. Rather they differed in the extent of their geopolitical aspirations, falling into two groups: the independence-minded republics (Estonia, Latvia, Lithuania, Moldova, Georgia, Armenia) and those which favoured some form of union (Ukraine, Belarus', Kazakhstan and the four Central Asian republics). The former Union republics aspired to the *nation state model*. At the forefront were the Baltic states of Estonia, Latvia and Lithuania. In these republics, nationalism quickly emerged to become separatist. It was based on people whose sense of national identity was already well developed before their forced incorporation into the Soviet Union in 1940 and where support for the cause of sovereign statehood owed much to the fact that all three were reclaiming the independence that they had enjoyed as nation states between 1918 and 1940. Moreover, the commitment to nation statehood was also strong in the Baltic republics precisely because these republics felt that their more developed economic base and past economic association with the West would ensure for their peoples higher living standards.

This contrasted with those republics which aspired to the *confederal model*. By this the republics would opt to remain part of the Soviet Union in return for which full sovereign powers would reside in the republics. The most supportive of this course were the five Union republics of Central Asia (including Kazakhstan). Unlike the Baltic republics, the peoples of Central Asia had no past experience of national independence to look back upon and indeed little sense of national identity before their incorporation into the Soviet Union. For these republics the appeal of statehood was also circumvented by economic realism. As the least developed of the Union republics and economically more dependent on Russia and the other Soviet states, Central Asians treated the idea of greater economic self-reliance with caution. Thus the Central Asian republics became independent states because, following declarations of sovereignty by Russia and the other Union republics, they had no choice but also to go down the path of nation statehood.

Russia and the new federal politics

The break-up of the Soviet Union is hardly exceptional among multi-ethnic empires in ending on an unplanned basis. What makes the Soviet experience distinctive is the metropole territory (Russia) becoming one of the first states to declare its sovereignty from the rest of the empire. This move in effect represented what Szporluk (1989) identifies as the victory in Russia of the 'nation builders' over that of 'the empire savers'. While the latter harboured a conception of Russia synonymous with the territorial borders of the Soviet Union, the nation builders see the future in a non-imperial Russia, one which has shed once and for all its legacy as a colonial power. Although voices in post-Soviet Russia can still be heard calling for the restoration of Russian rule in Ukraine, the Baltic states and Central Asia, for the nation builders the major task which confronts Russia is therefore not the re-annexation of its borderlands but rather that of retaining the territorial integrity of the new Russia. So in having shed its empire Russia is now confronted with how to manage a multi-ethnic polity in which nearly one in five of its citizens is non-Russian, most of whom are concentrated in their own territorial homelands.

In finding a way of managing its multi-ethnic character, Russia is unique among the post-Soviet states in opting for the retention of a federal structure. What in effect has been proclaimed into existence is 21 ethnically-designated republics (Fig. 3.1). In so doing, the new Russian state is faced with the mammoth task of convincing its own ethnorepublics[1] that such a federal arrangement is inextricably bound up with providing their peoples with a scale of territorial autonomy sufficient to satisfy their demands for local control over their political, economic and cultural affairs. Yet within the ethnorepublics the notion of federalism retains a pejorative meaning associated with a Soviet federal structure in which, as former autonomous republics, the ethnic regions enjoyed far less control over their local affairs than even the Union republics. Indeed they displayed many of the features of 'internal colonies' (Smith 1985):

1. *Limited degree of territorial autonomy*. Russia's autonomous republics (ASSRs) enjoyed even less autonomy over their local affairs than the Union republics. Their cultural rights were more heavily restricted (e.g. language rights, amenities for native schooling), with their peoples receiving much of their education in the Russian language (see Ch. 2). They also had fewer publications per head of population in their local languages than the non-Russian Union republic-based nationalities. In short they were subjected to a scale of political and cultural Russification far greater than that experienced by the Union republics. Such were the consequences of Russification that on the eve of the dissolution of the Soviet Empire nearly half of Karelians and a quarter of Komis, Udmurts, Mari, Karachayens and Cherkess spoke Russian as their native tongue (*Gosudarstvennyy komitet SSSR* 1991).

2. *Primary resource-based, core-dependent economies*. Although the centre practised a policy of regional specialization and economic inter-dependency throughout its republics, Russia's own ethnic regions were more characteristically dependent on primary resources thus displaying a far greater degree of limited and dependent development than the Union republics.

---

[1]The 21 republics within the Russian Federation, most of which are former autonomous republics (ASSRs). In addition, the Russian Federation currently contains two autonomous oblasts and ten autonomous okrugs, based also on ethnic criteria, as well as many 'non-autonomous' administrative units. However, the powers of autonomous units other than republics are limited and they have not presented the same scale of problems to the new federation.

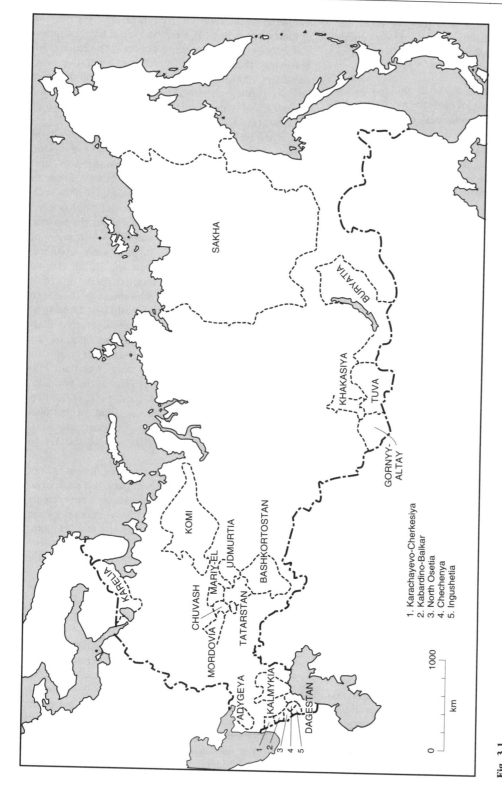

Fig. 3.1
Republics of the Russian Federation.

It was in effect tantamount to a form of economic colonialism in which the resources of the ethnic regions were heavily exploited, often with far-reaching environmental consequences.

3. *Limited social stratification.* As a consequence of their specialist form of development the native peoples of Russia's ethnic regions remained concentrated in primary economic activities with only a small native urban middle class. More specialist positions tended to be filled by Russian immigrants, a settlement process linked in part to the exploitation of local natural resources. Thus within each ethnic region, a detectable ethnic division of labour existed in which the native populations remained dependent on the primary economic sector, whereas Russians were more likely to be located in more specialist managerial, technical and administrative positions in the cities.

4. *Lower living standards.* Due to their general underdevelopment, living standards in Russia's ethnic regions were among the lowest in the Soviet Union, a legacy which has continued into the post-Soviet period. Thus below national average levels of income still persist in Chechenya, Ingushetia, North Osetia, Dagestan, Kabardino-Balkar, Mari-El, Mordovia and Kalmykia (*Izvestiya* 1992a).

For the architects of the new Russia, a more democratized federal arrangement was held up as a new beginning in relations between the centre and its own ethnoregions. For Russia's first president, Boris Yeltsin, such a redesignated federal arrangement held a number of attractions. Firstly, in the transition to securing Russian statehood (1990–91), Yeltsin, in supporting the autonomist demands of Russia's ethnoregions, saw a way of strengthening his own power base and of undermining Gorbachev's position as president of the Soviet Union. Thus Yeltsin openly encouraged the ethnic regions to declare themselves as sovereign entities with greater political powers. And in return, several of the ethnoregions boycotted Gorbachev's last-minute referendum in 1991 to keep the Soviet Union together. By so doing the ethnic regions signalled their support for Yeltsin and for the establishment of an independent Russia. Secondly, federalism offered a way of speeding up Moscow's commitment to the country's democratization. By 1990, Yeltsin argued that the only way that Russia could be effectively democratized was by restructuring the country 'from below' in which the various ethnic regions should have whatever powers they so wished (*Literaturnya gazeta* 1990). Finally, federalism was envisaged as a territorial strategy to prevent the break-up of Russia. By arguing that the ethnic regions

should be granted substantive powers over the running of their own regional affairs, the Yeltsin administration was effectively attempting to prevent the centre from becoming the focus of increasing frustration and animosity. Through adopting an accommodationist line, Yeltsin was in effect trying to deflect support for secession and so ensuring the territorial preservation of the new Russia.

The new federal idea was encapsulated in the Federal Treaty of 31 March 1992. Only two of the twenty-one republics, Tatarstan and Chechenya, refused to sign. Yet the Treaty, which redefined relations between the centre and the ethnic regions and which with modifications became law in December 1993, has been unable to satisfy some of the ethnic regions. In practice it falls short of the scale of home rule which many of the ethnic regions would like. Instead, some republics, most notably Chechenya, Tatarstan, Bashkortostan and Sakha (Yakutia), have adopted their own local constitutions, proclaiming the supremacy of their own republican laws over Russian federal laws. Striking at the heart of this tension is the omission from draft federal constitutional proposals of the right of the ethnic republics to secede, if they so wish, from the Federation, a right which some claim was built into the original 1992 Federal Treaty. This, as the republics see it, is the abrogation of a basic right of nations to practise, if they so wish, the right of national self-determination, a right that, in theory at least, was even available during Soviet rule to the Union republics.

The controversy over sovereignty is, however, not confined to the right of an ethnoregion, if it so desires, to constitute a nation state. The nature of sovereignty also raises other issues to do with the federal nature of relations between the centre and the ethnoregions. These include: the relationship between federalism and democracy, the nature of centre–local economic relations, and issues of cultural autonomy.

While the new Russia has declared its commitment to democracy through a more decentralized federal system of government, the republics criticize Moscow for continuing to practise the 'dictate of the centre'. The fear is that Russia, given its historically-centralizing impulse, is again sliding back towards centralized authoritarianism (*Rossiyskaya gazeta* 1993b). Indeed Yeltsin's dismissal of the Russian parliament in October 1993 and its replacement with rule by a three-month period of presidential decree is interpreted in the ethnoregions as yet another stage towards the recentralization of political power, even although this was followed in December 1993 by fresh federal elections, as promised. Moreover, there is concern over constitutional statements which proclaimed that in the new Russia what should take precedence 'is the individual and his [sic] inalienable rights' and that collective rights, such as ethnic rights, are of secondary importance (*Rossiyskaya gazeta* 1993c). The ethnoregions therefore remain suspicious of Yeltsin's commitment to a more democratized Russia in which both their participation in decision making and their rights as ethnic groupings will be guaranteed.

Yet it would be mistaken to assume that within their own localities, democracy is necessarily high on the agenda of the ethnic republics. It is incorrect to claim, as experiences elsewhere show, that the desire for local accountability automatically leads to greater local democracy. There is no basis in political theory for claiming that smaller territorial units would be more hospitable to democratic politics. In Russia and its regions, where local civil society is weak and where there is no tradition of pluralist democracy, this is likely to be especially problematic. In some republics, notably Mordovia and Kalmykia, where Moscow has in effect lost control over local affairs and where the regions are pursuing their own economic and political policies, authoritarian trends are already detectable. In these *de facto* sovereign states, conservative-minded leaders, many of whom were in power during the late Soviet period, justify semi-authoritarian rule by claiming that it provides both social stability and economic direction in times of political uncertainty.

The issue of local economic autonomy and fiscal accountability to the centre is also of concern in the ethnoregions. Most want to pursue their own local paths to development. Yet adjusting from being a centrally planned economy in which each region was told what to produce and where product markets were guaranteed as part of central planning has been extremely painful. The ethnoregions envisage that the most effective way of realizing this transition to market exchange is by implementing policies sensitive to local economic conditions and that means greater local control. Such autonomy would also mean having greater control over their own fiscal affairs. Not surprisingly, the ethnoregions resent having to contribute to a federal budget that necessitates cutting back on local public spending, be it on economic investment or social welfare. So in taking greater control over their local economies, the republics see a way of not only using their own resources to the advantage of building up their local economies but also of pursuing more effectively their transition to the market. One republic at the forefront is the North Caucasian republic of Kalmykia. In order to escape from being one of the poorest and least economically viable of the Russia's republics, Kalmykia has introduced a far-reaching programme of market liberalization designed to restructure its economy from

its specialist, export-dependency on agriculture and other primary products to one based on greater economic diversification through establishing a range of competitive manufacturing-based industries. It is a programme for economic renewal which plans to rely on outside capital investment through tax-free incentives and which is designed to end the republic's dependency on economic subsidies from Moscow (Tolz 1993). It is, however, a particular developmental strategy which combines a commitment to rapid economic growth with the idea of the strong local state, one in which securing economic growth takes precedence over promoting pluralist democracy.

Finally, there is the issue of cultural autonomy. The ethnic regions want to reclaim their local identities through promoting their own national languages, national cultures and religions. Yet the ethnorepublics fear that such local policies are at odds with a federal structure in which a strong and regionally-insensitive centre may be keen to re-establish its own cultural dominance through promoting Russification. This is especially pertinent given the sizeable demographic presence of Russians in the ethnoregions and the fear that local Russians may again be used as agents of centrally-initiated policies of Russification. In promoting their own national cultures, however, many ethnoregions have also been accused by the centre of initiating an overly nationalistic form of cultural dominance. This has fuelled tensions between local Russians and natives and as a consequence many Russians, fearful of their future, have left (*Izvestiya* 1992b). It is a fear not without substance. A handful of republics have even gone so far as to propose a definition of local citizenship in which the right to participate in republic elections would be open only to those who are of the local nationality (*Literaturnaya gazeta* 1992). In some republics, where the dominant local culture is Islamic, the fear of the establishment of a fundamentalist state is also causing concern among local Russians, notably in Muslim Tatarstan.

The federal idea is in many respects treated with suspicion in the ethnoregions and in the minds of many is still associated with a form of Soviet rule which was both centralist and largely insensitive to the ethnoregions. According to Tarlton's (1965) seminal essay on federalism, in polities where each locality differs in terms of a range of criteria from the national state in general (e.g. ethnicity, language, economic development), 'relieving the tensions and discord . . . requires not further recognition of diversity and their protection in the complicated processes of ever increasing federalization, but rather increased co-ordination and coercion from the centralizing authorities in the system' (p. 874). Tarlton was clearly concerned, as indeed Moscow is, that 'when diversity predominates, the "secessionist potential" of the system is high and unity would require control to overcome disruptive, centrifugal tendencies and forces' (p. 873). Many centralists in Moscow have already called for a 'harsh federation', with greater dependence of the regions on the centre, insisting that without strong power it would be impossible to effect Russia's successful transition to the market (Sakwa 1993). But centre-building strategies to 'contain ethnoregionalism' are already apparent. These have involved re-emphasizing the need for bilingualism within the ethnoregions through playing up the importance of the Russian language as a factor in inter-regional communication to using force to resecure federal control over wayward republics (e.g. Chechenya) and in arming local Russian-based pro-federal organizations against nationalist-secessionist movements in the North Caucasus.

Such centralist strategies of coping with an ethno-regionally diverse federal union have only added fuel to the separatist cause. For more radically-minded regional nationalists, the only solution is to break with Russia geopolitically. Yet there are a number of factors which militate against secessionist nationalism.

Firstly, there is the ethnodemographic composition of the ethnoregions. Russians constitute a majority of the population in six republics, notably in Buryatia (72 per cent), Karelia (71 per cent), Mordovia (60 per cent), Udmurtia (58.3 per cent), Komi (57 per cent), Sakha (50 per cent), and a sizeable minority in all the others. Although many local Russians have supported local autonomy, they are more likely to identify with what they regard as their homeland, Russia, and support the preservation of a Russian Federation. Not to do so would, in the case of a number of republics, transform their status from being part of the majority nation to that of a minority whose ethnic rights and status are more likely to be threatened. Thus given the sizeable if not majoritative status of Russians in a number of these republics, any local referendum on secession is likely to result in support for remaining part of the Russian Federation.

Secondly, there is the question of economic viability. Despite the resource-endowed richness of many of these republics, such as Sakha which accounts for most of Russia's gold and diamond output and oil-rich Tatarstan and Bashkortostan, they are unlikely to be transformed into the 'Kuwaits of northern Eurasia' overnight. As a consequence of internal colonialism, their economies remain inextricably dependent on that of Russia and in some cases, notably Tatarstan, this is reinforced by their landlocked status. At the other end of the economic spectrum are the poor republics like Kalmykia and Tuva which still depend on large economic subsidies from Moscow. The nature of this economic dependence and the tangible benefits of remaining part of a larger and integrated trading community would outweigh the greater economic uncertainty of independent statehood.

Finally, it is important not to underestimate the centuries-long impact on ethnoregional identities of being part of Russia. One consequence has been the structural assimilation of many urbanized and more upwardly mobile native peoples into adopting many of the cultural values of Russians. Such assimilated peoples are less likely to be supportive of local independence. Even for the majority who retain a strong sense of ethnoregional identity, 70 years of Soviet rule have also had an integrative impact in which self-identification of being Chechen, Sakha or Buryat does not necessarily conflict with remaining part of a multicultural Russia.

The non-Russian states, citizenship and minority rights

In the fourteen non-Russian states, nationalism is also in part bound up with state building. For the new national élites in power, what has become central in their concern with prioritizing nation building as part of their project of state building is the quest for national congruence. Williams has defined such a feature of state building as the attempt to make both national community and territorial state into coextensive entities (Williams 1989:196). In the case of the post-Soviet states, this has taken two particular forms. Firstly, it has manifested itself in extra-territorial claims based on a desire to redefine state boundaries to incorporate all members of the national community. One of the most contentious and violent concerns the disputed territory of Nagorno-Karabakh in the Transcaucasus. Here the peoples of the predominantly Armenian enclave of Azerbaijan have been supported by the Armenian state in their wish to be reunited with their co-nationals in the same polity. For Armenians, Nagorno-Karabakh is considered historically a part of ancient Armenia, and Stalin's decision in 1923 to include it within Muslim Azerbaijan was judged as an affront to Christian Armenians, particularly as it was awarded to a predominantly Muslim country.

Secondly, the quest for national congruence also involves ensuring a central place for the national community within the cultural, economic and political life of its own territorial state. Unlike Russia, which officially declared itself to be a multi-ethnic polity, the other successor states have been proclaimed into existence as national states. Yet, as all the non-Russian states contain sizeable ethnic minorities, none can legitimately claim to be national states (Table 3.1).

Table 3.1 The three biggest ethnic groups in each of the post-Soviet republics, 1989

Republic	Percentage titular nationality	Percentage of of second nationality		Percentage of third nationality	
Russia	81.5	3.8	(Tatars)	3.0	(Ukrainians)
Ukraine	72.7	22.1	(Russians	0.9	(Jews)
Belarus'	77.9	13.2	(Russians)	4.1	(Poles)
Uzbekistan	71.4	8.3	(Russians)	4.7	(Tajiks)
Kazakhstan	39.7	37.8	(Russians)	5.8	(Germans)
Georgia	70.1	8.1	(Armenians)	6.3	(Russians)
Azerbaijan	82.7	5.6	(Russians)	5.6	(Armenians)
Lithuania	79.6	9.4	(Russians)	7.0	(Poles)
Moldova	64.5	13.8	(Ukrainians)	13.0	(Russians)
Latvia	52.0	34.0	(Russians)	4.5	(Belorussians)
Kyrgyzstan	52.4	21.5	(Russians)	12.9	(Uzbeks)
Tajikistan	62.3	23.5	(Uzbeks)	7.6	(Russians)
Armenia	93.3	2.6	(Azeris)	1.7	(Kurds)
Turkmenistan	72.0	9.5	(Russians)	9.0	(Uzbeks)
Estonia	61.5	30.3	(Russians)	3.1	(Ukrainians)

Source: 1989 Census of Nationality.

Indeed there are some 40 million people who now find themselves ethnic minorities in these new states. This quest to ensure a privileged place for the dominant nation has prompted the new states to be labelled 'ethnocratic' because of the advantages, if only linguistic, which they now enjoy (Sheehy 1993). What, therefore, becomes crucial for ethnic relations in these new states is the extent to which the idea of the national state can coexist with the reality of their multi-ethnic composition.

The minority rights question has been especially contentious concerning the 25 million Russians who are scattered throughout non-Russian states and who comprise over half of all ethnic minorities (Fig. 3.2). They constitute a sizeable minority in more or less every state, particularly in Ukraine (22 per cent), Estonia (30 per cent), Latvia (34 per cent) and in Kazakhstan (38 per cent). Although many are descendants of settlers who moved into Russia's borderlands in pre-Soviet times, a large number constitute more recent waves of migrants who moved in during Soviet rule. What has made the issue of the Russian minorities especially critical is their associational link with the cultural heartland of the former Soviet Empire. During Soviet rule, Russians enjoyed extra-territorial privileges as a result of their nationality status, including their own language-based schools, cultural facilities and access to prestigious employment, all of which carried no implication to learn the local languages. Not surprisingly, as noted in

Chapter 2, knowledge among Russians of the new state languages is low, ranging from 5 per cent in Central Asia to 20 per cent in the Baltic states. Hence of all the nationalities, the Russians living in the non-Russian borderlands had most to lose from the end of the Soviet Empire. Moreover in the mindset of the local populations, Russians are still perceived as 'colonizers', bound up historically with Soviet rule and oppression. As international relations between some of these new states and Russia have deteriorated, and as the fear remains among many in the borderlands that Russia may again become involved in empire building, Russian minorities are also judged as a possible geopolitical security risk.

Most of the new states opted formally to accommodate their ethnic minorities, including their Russian communities, by adopting a territorial definition of citizenship. By this formula, all those resident on national territory at the time of independence were offered full political membership of the new polity, including the right to participate in elections. Only two republics, Estonia and Latvia, opted for an alternative and less inclusive definition, not far removed from an ethnic model of citizenship. Consequently only those who were citizens or descendants of citizens during the former period of independent statehood (1918–40) were deemed to have the right to full membership of the sovereign community. For residents and their descendants who arrived during Soviet rule, the majority of whom are Russians, both a length of

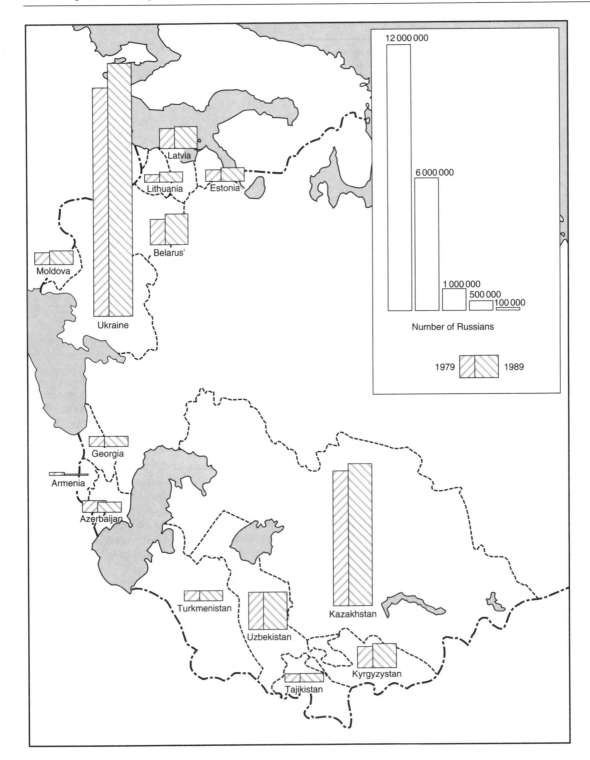

Fig. 3.2
The Russian diaspora in the non-Russian states, 1979 and 1989.

residency qualification (two years in Estonia and ten years in Latvia) and an ability to communicate in the new state language are required before an applicant is considered for citizenship.

That Estonia and Latvia should have edged towards a more ethnic as opposed to a territorial definition of citizenship was bound up with a fear that because of ethnodemographics their native peoples would not be in full control of their polities. Although all the non-Russian republics (with the exception of Georgia and Armenia) received large influxes of Russians during Soviet rule it was only in Estonia and Latvia that such a massive immigration, in tandem with their exceptionally low birth rates, raised concern about the titular nations becoming minorities in their own homelands. In Latvia, the native population had declined from 77 per cent in 1939 to 52 per cent by 1989 and in Estonia from 88.2 per cent to 61.5 per cent. During the same period, the Russian share had increased from about one-tenth to about one-third in both republics. It was also a scale of migration which, in contrast to the other non-Russian republics, had not eased up in the 1970s and 1980s.

This, however, does not mean that relations between Russians and the titular nations are necessarily more harmonious elsewhere. Despite more inclusive citizenship laws, state ethnocratic practices associated with language rights, public sector employment and housing have led to fears among many Russians of their status being reduced to a new ethnic underclass. Yet despite being granted the right to citizenship, many Russians have been reluctant to become citizens of the new states. This 'opting out' is also in part a product of Russia adopting an extra-territorial definition of citizenship which offers all those resident in the former territories of the Soviet Union the opportunity to become citizens of Russia. Russians in the borderland states are therefore faced with a dilemma. On the one hand they do not want to sever ties with their cultural heartland, particularly if things become so desperate in their place of residence that returning to Russia may be necessary. On the other hand, by not taking up the opportunity of citizenship in their place of residence they would risk becoming marginalized. However, in those states where the possibility of citizenship exists, most Russians have chosen to become full members of the new polity.

The geopolitics of ethnoregionalism

In a number of the non-Russian states Russians are concentrated in particular borderland regions, raising the possibility of their region seceding from the state of which they form a part and reincorporation into a Greater Russia. In ethnoregions such as eastern Ukraine (Donbass), north-east Estonia and northern Kazakhstan, the sense of being part of a Russian community is strengthened not only by their contiguous location to their cultural heartland but also by their concentration in localities where the Russian language and way of life predominate. It is particularly within these ethnoregions that their future geopolitical relationship both with Russia and the polity of which they are a part has been most hotly contested.

Underlying the ethno-politics of these Russian communities is their cultural identity. On the one hand there is a sense of Russianness which associates itself with the cultural heartland of Russia. Here neither migration nor membership of a new polity has had a sufficient impact in altering a cultural affinity with Russia. Indeed, in many respects, finding themselves as outsiders 'abroad' may reinforce this sense of identity with Russia proper. On the other hand, there is a sense of Russianness which is associated with the local ethnoregional community. There is evidence to suggest that migration from Russia has had a transformative impact in detaching Russians from identifying with their cultural heartland which has led to adopting their own particular values and customs. This is also linked to a degree of acculturation into the dominant local culture, as evidenced by knowledge of the titular language, inter-ethnic marriage and the adoption of many local values and traditions. It is especially long-settled Russians who have been susceptible to acquiring an identity distinct from their co-nationals in Russia (Aasland 1995).

However, it is only under certain conditions that such differing scales of geocultural identity are likely to be translated into specific geopolitical demands. Thus at one end of the spectrum we would expect that Russians with a stronger cultural association with Russia are more likely to support their region's geopolitical secession than those whose cultural links with Russia are weaker. At the other end of the spectrum we would expect that Russians with a stronger sense of affinity with their place of residence are more likely, other things being equal, to accept the reality of their situational context and seek to play an integrative role within the new state. On this basis, by juxtaposing geocultural identities with geopolitical aspirations, we can arrive at the following three-fold typology of Russian ethnoregionalism in the non-Russian states (Fig. 3.3).

First, there is what we can call an *irredentist nationalism*. Here the identity of the ethnoregional community is indistinguishable from the cultural heartland. Members seek to secede from the polity of which they are a part and be reunited with their

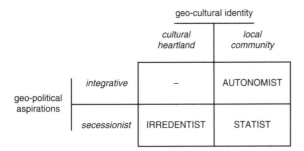

Fig. 3.3
Russian ethno-regionalism in the borderland states of the former Soviet Union.

co-nationals within the new Russia. We find such demands evident within a minority of the population of north-east Estonia, the Donbass region of Ukraine and northern Kazakhstan. It is associated particularly with recent migrants whose identity with Russia is strongest and who are least familiar with the new state language. In a survey undertaken in February 1993 of the predominantly Russian city of Narva in north-east Estonia, it was found that of the one in five Russians who supported secession, most were recent arrivals whose knowledge of Estonian was invariably non-existent and whose sense of identity with the cultural heartland was strong (Smith 1994).

Secondly, we can identify a desire for *regional autonomy*. Here the goal is to secure ethnoregional self-government within the new polity. It is associated particularly with those Russians whose sense of identity is more integrated with the polity in which they reside. Such goals include securing cultural autonomy for their own languages, local schools and other cultural facilities, fiscal control over their own local budgets, and fair representation in both local and national government. This is at present the prevailing view in both north-east Estonia and in eastern Ukraine. In the latter region, where Russian speakers make up between one-half and two-thirds of the population, concern focuses on both cultural and economic matters. The Russian-dominated regional organizations which have sprung up in the region since Ukraine's independence in 1991 have signalled their resolute opposition to proposals to Ukrainianize local education as well as plans to cut back on state subsidies to the region's inefficient smokestack industries, fearing large-scale local unemployment (Wilson 1993). In both eastern Ukraine and north-east Estonia, however, the situation is so fluid that if autonomist demands are not accommodated then nationalism could take on an increasingly irredentist form.

Finally, we can identify a *statist nationalism*. By this we mean that a Russian minority neither wants to

remain within the polity of which it is a part or be incorporated into Russia but rather seeks to secure a nation state of its own. The only example of this type are the Russians in the Transdniester district of Moldova who have set up their own sovereign enclave. Its geopolitical future, however, is uncertain not least because of Moldova's opposition to what its government sees as loss of an integral part of Moldovan territory. Although the cultural identity of Moldovan Russians is strongly associated with Russia, the practicality of their non-contiguous status dictates an alternative course to that of becoming part of Russia. Should a Transdniestrian state succeed it is probable that Russians in this state would acquire a cultural identity of their own, separate from that of Russia (Kolsto *et al.* 1993).

Conclusions

The transition from empire to nation states has not only restructured political space, but also the political space occupied by various national and ethnic groups. In the process, it has heightened national and ethnic tensions as various ethnic communities seek to redefine their relationship with each other and with the new sovereign spaces in which they now find themselves. What, however, the experiences of newly independent states emerging from colonial rule in the Third World show is that without providing disenchanted ethnic minorities with both ethnic and citizenship rights, there is little prospect of social harmony. The extent to which the post-Soviet states are willing to provide their citizens with full and equal participation irrespective of their cultural differences will be crucial in determining whether ethnic coexistence or violence becomes the central feature of statehood.

References

Aasland A 1995 The Russian diaspora. In Smith G. (ed.) *National and Ethnic Relations in the Post-Soviet States*, 2nd edn. Longman, London
Arutyunyan Yo V *et al.* 1992 *Russkiye. Etnosotsiologischeskiye ocherki.* Nauka, Moscow
Gosudarstvennyy komitet SSSR po statistike soobshchayet 1991 *Natsional'nyy sostav naseleniya* (Moscow), vol. 11, pp. 3–5
Izvestiya, 1992(a), 13 February
Izvestiya, 1992(b), 4 July: 3
Izvestiya, 1993, 17 May
Kolsto P, Edemsky A, Kalashikova N 1993 The Dniester conflict: between irredentism and separatism. *Europe–Asia Studies* **45**(6): 973–1000
Literaturnaya gazeta, 1990, 15 August
Literaturnaya gazeta 1992, 21 May

Moskovskiye novosti, 1992, (41), October: 2

Pravda, 1989, 17 August

Rossiyskaya gazeta 1993(a), 19 May

Rossiyskaya gazeta 1993(b), 26 May

Rossiyskaya gazeta, 1993(c), 13 June

Rossiyskiye vesti, 1991, 15 June

Sakwa R 1993 *Russia: Politics and Society*. Routledge, London

Sheehy A (1993) The CIS: a shaky edifice. *RFL/RL Research Report* (1): 38–43.

Smith G 1985 Nationalism, regionalism and the state. *Environment and Planning C. Government and Policy* 3(1): 1–9

Smith G 1989 *Planned Development in the Socialist World*. Cambridge University Press, Cambridge

Smith G 1990a The Soviet federation: from corporatist to crisis politics. In Chisholm M, Smith D (eds) *Spared Space, Divided Space. Essays on Conflict and Territorial Organisation*. Unwin Hyman, London, pp. 84–105

Smith G (ed.) 1990b *The Nationalities Question in the Soviet Union*. Longman, London

Smith G 1994 Statehood, ethnic relations and citizenship. In Smith G (ed.) *The Baltic States. The national self-determination of Estonia, Latvia and Lithuania*. Macmillan, London, 181–205

Szporluk R (1989) Dilemmas of Russian nationalism. *Problems of Communism* **38**(4): 15–35

Tarlton C 1965 Symmetry and asymmetry as elements of federalism: a theoretical speculation. *The Journal of Politics* **27**: 861–74

Tolz V (1993) Russia's Kalmyk Republic follows its own course. *RFE/RL. Research Report* **2**(23): 38–43

Williams C 1989 The question of national congruence. In Johnston R, Taylor P (eds) *A World in Crisis?* Basil Blackwell, Oxford, pp. 196–230

Wilson A 1993 The growing challenge to Kiev in the Donbass. *RFE/RL. Research Report* **2**(33): 8–13

4

Industrial policy and location

Robert N. North and Denis J.B. Shaw

Industry was central to Soviet economic development and has a major role to play in the economies of the newly independent republics. This chapter examines the development and spatial distribution of Soviet industry and their implications for the USSR's successor states. The last section considers the changes which are now occurring in those states and what such changes may mean for their future industrial geographies.

Industrial policy and location up to 1953

Soviet industrial output in 1953 was only one-tenth of what it was at the end of the Soviet period. Nevertheless, if we wish to understand modern industrial location there are good reasons for paying attention to the 1953 situation and how it came about. The first is the importance of inertia. In any country, the locations first chosen for industry acquire advantages which help to preserve their dominance. They include: firstly, a reserve of invested capital, in industrial plant, utilities, housing and community facilities, which can often be added to or adapted more cheaply than opening up new locations; secondly, a transport system which tends to preserve existing patterns of accessibility and centrality, especially if railways are the principal mode; and, thirdly, a pool of technical and managerial skills.

In the former Soviet Union, despite reiterated government intentions to spread industry more evenly through the country, inertia was at least as important as in any Western country. There are particular reasons for this. As noted in Chapter 2, during most of the Soviet era, industry was managed by ministries, responsible for both regulation and production in such broad sectors as iron and steel, heavy engineering, and pulp and paper. Until late in the period, their main task was to maximize output with minimal investment. Such measures of efficiency as cost minimization were

subsidiary, and in any case applied to whole ministries rather than individual enterprises. The ministries therefore had every incentive to keep old facilities running, provided that they did not affect the ministries' overall performance indices too drastically, and to seek permission to build new facilities in established locations, where they could be brought on stream most rapidly. The Soviet Union, therefore, largely avoided the problem of redundant, half-abandoned industrial regions which has plagued western Europe. The corollary is that few major new industrial regions emerged during Soviet times. At the end of the Soviet period, all but one of the leading manufacturing regions were still those of 1953, or even in most cases 1914. The situation was exacerbated by the almost unvarying primacy of national over regional development policies. To the extent that industrial investment in outlying regions was seen as an expensive luxury detracting from fast national growth – rather than a strategic or political necessity – it was not favoured.

That is not to say that there were no changes at all in industrial location after 1914. A second reason for examining the years before 1953 is that the most distinctive changes in Soviet times occurred under Stalin. For example, heavy industry received much more investment than light industry and was used as a development leader in new regions. Therefore, it came to reflect new locational influences (Fig. 4.1), while light industry, especially textiles, retained pre-revolutionary characteristics (Figs. 4.2 and 4.3). Also, since transport investment focused on a few trunk railways, manufacturing tended towards a nodal distribution. In other words, a good deal of what appears to Western eyes to be 'different' in Soviet industrial location stems from the Stalin period.

A third reason for examining the period before 1953 is to compare it with more recent times and demonstrate that Soviet locational policy was far from consistent. For most Western countries that point

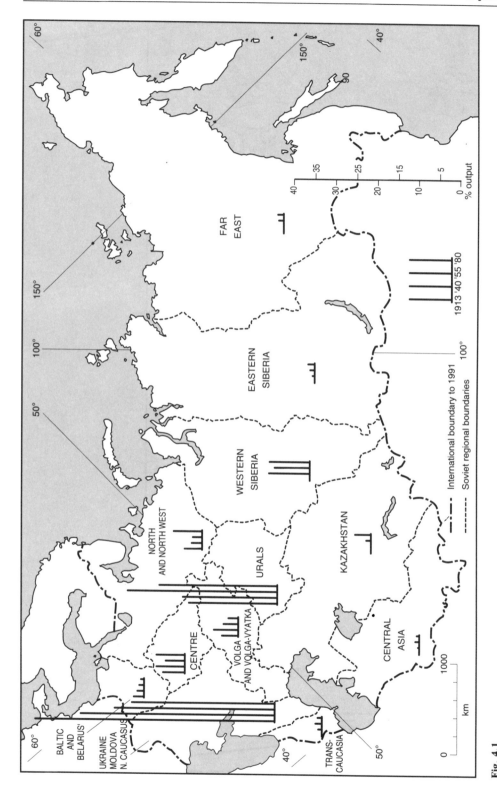

Fig. 4.1
Steel ingot production: regional shares, 1913–80.

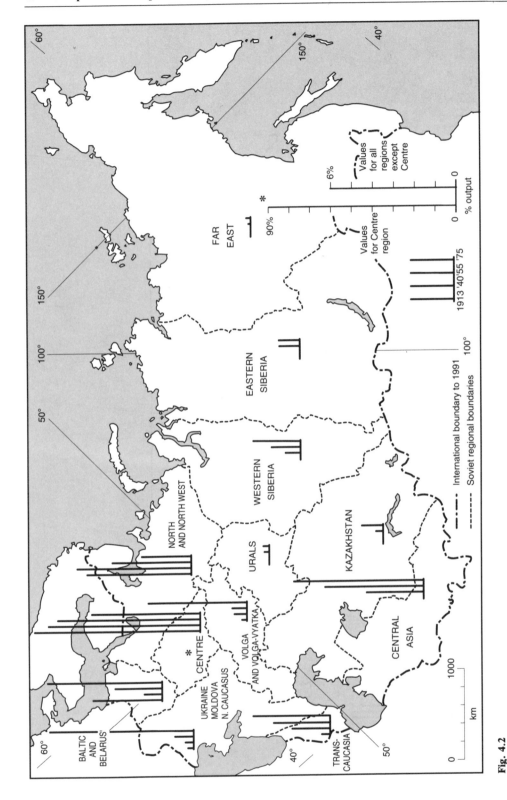

Fig. 4.2
Cotton cloth: regional shares of output, 1913–75.

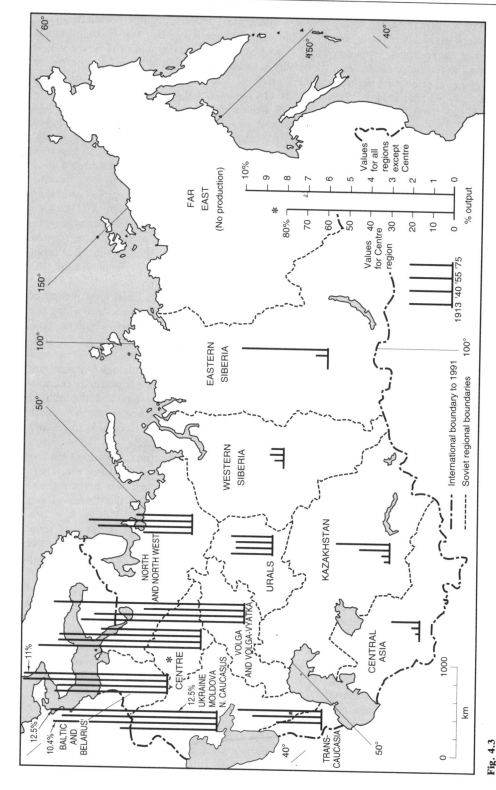

Fig. 4.3
Woollen cloth: regional shares of output, 1913–75.

might seem too obvious to be worth making. However, Soviet economic geographers generally claimed that the territorial organization of the economy developed during Soviet times according to certain economic laws (which might be seen as planning principles) peculiar to socialism. The list was modified over time: a 1985 text (Khrushchev and Nikol'skiy 1985:23) presented it as follows:

1. The planned and proportional distribution of production throughout the country in order to make the most rational use of natural, material and labour resources and most fully satisfy the needs of the population.
2. The location of production close to sources of raw materials, fuel and energy resources, available infrastructure, labour resources, and consuming regions.
3. The elimination of socio-economic and cultural differences, and differences in living conditions, between town and country, and the fostering of the integration of agriculture and industry.
4. The development of Union republics and economic regions to achieve rational levels of economic specialization on the one hand, and rounded development on the other.
5. The formation of Territorial Production Complexes (TPCs) as the bases of economic regions.
6. The specialization of regional TPCs in lines of production requiring minimal material and labour inputs under given economic and natural conditions.
7. Equalization of the level of development of economic regions.
8. The distribution of production to take account of cooperation among countries of the world socialist system and to promote socialist economic integration.

Clearly the principles were not so precise as to indicate specific spatial patterns of development. Whether production was to be located close to raw materials or consuming regions when they did not coincide, and how close, presumably depended on circumstances. How rapid a spread of economic activity constituted acceptable progress towards equalization of regional development must have been a matter of judgement. Nevertheless the principles were regularly cited as an ultimate reference point in socialist industrial location. However, despite the consistency they were claimed to represent, many changes since 1953 appeared to be reactions against what was created under Stalin.

The Soviet inheritance from tsarist Russia

Before the First World War the Russian Empire exported primary and processed goods such as wheat,

dairy products and petroleum, and imported two-thirds of the manufactures it needed. But it could already be described as a second-tier industrial power with a healthy growth rate. It was the world's fifth-largest steel producer and fourth-largest producer of cotton goods. Overall industrial production was only one-fifth that of Britain and one-fifteenth that of the United States, but it was growing faster than in any established industrial country except the United States, Japan and Germany.

Pre-revolutionary industry was concentrated in a few areas. If we include only what later became part of the Soviet Union, five regions accounted for 80 per cent of the industrial output (Fig. 4.4). That centred on Moscow was the oldest and largest. Local entrepreneurs had built a textile industry, first using flax, the local raw material, and later wool and cotton. The cotton came originally from the United States, but by 1914 from Central Asia. From textiles the region followed a classical sequence of industrial development, working backwards along the production chain through textile machinery to machine tools and primary steel production, and through dyes to basic chemicals. The Moscow region was well favoured for industrial growth. Although St Petersburg became the capital early in the eighteenth century, Moscow remained the largest city and the hub of communications. Indeed until late in the nineteenth century it was one of very few cities able to reach a wider-than-local market effectively.

Industry in Ukraine, by contrast, was almost entirely a creation of the last 50 years of the tsarist era. There were two components. The first was a mining and metallurgical complex based on the Donets Basin coalfield (the Donbass) and the Krivoy Rog iron ore deposit, joined by rail over a distance of some 350 kilometres (Fig. 4.5). By 1900 iron and steel works were located at both ends of the axis, as well as on the River Dnepr, where the water supply was much better than at Krivoy Rog, and on the coast of the Sea of Azov. The government encouraged foreign investment to speed industrial growth and provide modern technology, as indeed it did elsewhere.

The second component of Ukrainian industry was an agricultural processing complex in the western Ukraine. Kiev was the centre of sugar refining based on local beet, and tropical imports were processed at Odessa.

The third industrial region comprised the Baltic ports from Riga to St Petersburg. Industries were based on imported raw materials – including coal and iron, so poor were transport links to domestic resources – and, in St Petersburg, on the presence of the government, the court, and a substantial local market.

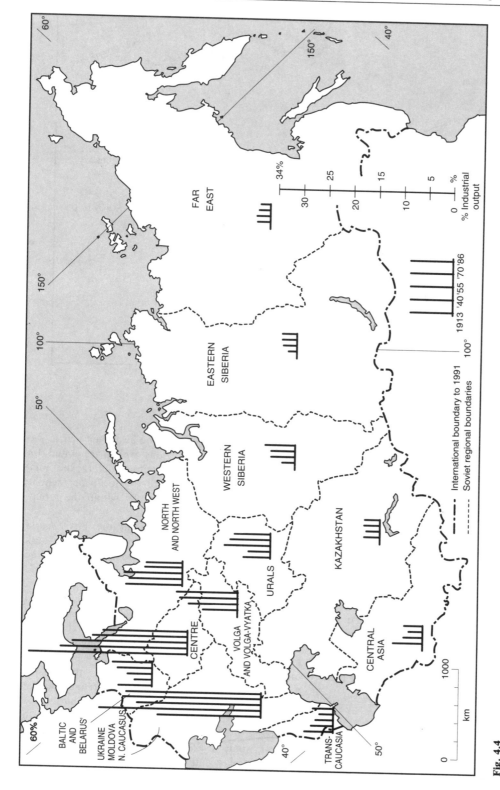

Fig. 4.4
Industrial output, 1913–86: estimated regional shares.

Fig. 4.5
The southern Ukrainian industrial region.

The two remaining industrial regions were Transcaucasia (basically Baku and Chiatura). The former depended on petroleum and manganese, mostly exported, and west European capital. The latter depended on iron ore, timber, and domestic capital. The charcoal-based iron industry of the Urals had been founded in the time of Peter the Great. It produced more pig-iron than Britain in the mideighteenth century, but at the end of the nineteenth century it could not compete with its coal-based rival in Ukraine.

Outside the five regions, industry was scattered through the Empire, but there was very little east of the Urals or in Central Asia. European Russia contained over 80 per cent of the population, and the government saw the eastern regions as additional markets to be preserved for European Russian producers rather than as potential locations for new industry.

Soviet policies and achievements to 1953

Considering that overall industrial output grew nearly twenty-fold from 1913 to 1953 by Soviet estimate, and that declared policy was to spread industry more

widely through the country, the locational inertia was remarkable. The five regions which accounted for 80 per cent of industrial output before the First World War still accounted for 60 to 70 per cent in 1953. However, there were shifts within the group, and new regions did emerge.

Destruction was very heavy in the First World War and in the civil war, and restoration took until the late 1920s. Thereafter we can identify three periods: before, during, and after the Second World War. Some factors were constant throughout. Firstly, the government desired fast economic growth based on industrialization. However, secondly, trade-based development was considered unreliable: the 1930s world economic depression had frustrated attempts to rebuild agricultural exports as a means to pay for technological imports. In any case, the outside world was considered hostile and an unreliable source of technology or strategic raw materials. Thirdly, the government wanted to spread development both to new regions and to small towns within each region. However, only to a limited extent was it prepared to compromise fast national growth to that end. Finally, strategic considerations did support industrial dispersal, and especially a reduced dependence on the western regions, which both the Napoleonic Wars and First World War had shown to be vulnerable.

The first two factors were reflected in an investment policy designed to build up an independent industrial economy, supported by military power, as fast as possible. Priority was given to a limited range of industries which could provide materials to build up other industries and the armed forces, especially mining, iron and steel, and mechanical engineering. Consumer goods, and even other heavy industries like petrochemicals and aluminium, were relatively neglected.

This set of factors and consequent policies are clearly reflected in the changes in the spatial distribution of industry which took place before the Second World War (Fig. 4.4). Among the established industrial regions, Ukraine enhanced its position dramatically, being the region able to expand heavy industry most rapidly. The Urals region took longer, since it had to replace outdated ironworks and shift to using coal. This process is usually associated with the Urals–Kuznetsk Combine, initiated under Stalin in 1928, but in fact by 1928 a quarter or more of Ural iron output was already based on coal from the Kuzbass in West Siberia. The Combine brought the building of full-cycle iron and steel works to the Urals and the Kuzbass, and an exchange of coal for iron ore along the 2000-kilometre railway between them. By 1950 the Urals had nearly doubled its share of national iron and steel output.

Established regions not already focused on high-priority heavy industries benefited less from the new policies. Even the Moscow region was only a partial exception. It was well away from the frontiers and had once again become the seat of government. It had research facilities and a pool of skilled labour, and it could expand production quickly in many industries. Nevertheless, its share of national output gradually declined. Industrial regions focused on ports suffered greater relative decline. Leningrad (as St Petersburg was known between 1924 and 1990) grew less than half as fast as Moscow, and Odessa fell from fifth to fourteenth largest city in the country. Finally, Transcaucasia was considered vulnerable and its hydrocarbon resources inadequate to fuel national growth. Other petroleum resources were not known at the time, while coal reserves were known to be enormous, so coal became the prime national energy source.

Among newer regions, industrial investment before the Second World War focused on West Siberia and Kazakhstan because of their resources of coal and non-ferrous metals. Whether the minerals should provide a basis for manufacturing, in a region of limited markets, or simply be extracted, and if so on what scale, had

been a topic of debate in the 1920s. A Siberian lobby advocating heavy investment had been opposed by a Ukrainian lobby promising faster national growth if investment were concentrated there. The government decided in favour of Siberia to the extent of building one full-cycle steelworks at the eastern end of the Urals–Kuznetsk Combine. Industrial growth rates in Siberia and Kazakhstan were impressive, but there was in fact much more investment further west.

Despite the attention usually given the Combine in Soviet and Western writings, it proved very expensive to operate and was soon de-emphasized. Karaganda coal, though of relatively poor quality, partly replaced Kuzbass coal in the Urals as soon as a railway could be built, because it was nearer. Similarly, Kuzbass reliance on iron ore from the Urals was reduced as soon as deposits south of the Kuzbass could be exploited.

The Second World War disrupted the emerging spatial pattern of industrial growth. In particular, the relative status of the eastern regions was enhanced by the eastward evacuation of several hundred factories in 1941, and by accelerated development to offset the loss of capacity to the Germans in the west. The former measure also gave the eastern regions a far greater variety of industries, especially in mechanical engineering, than they would have acquired by 1945 if prewar trends had continued. The impact was not confined to the war years: many factories remained after 1945. The regions most affected were along the Trans-Siberian railway from the Volga region, the biggest recipient, to West Siberia. Central Asia, with less easy access to the war zone, received fewer factories.

The postwar years were largely devoted to rebuilding in the west, but even by 1950 all the formerly German-occupied regions were still below their 1940 share of national industrial output. The only exception was the Baltic states, which in 1940 had only just been regained by the Soviet Union.

At Stalin's death, therefore, the Soviet industrial location pattern reflected three sets of superimposed influences, namely those of the tsarist era, the prewar five-year plans, and the Second World War. In the last two periods political and strategic goals were clearly important. A more complex question, to which we return later, is the role of economic calculations in industrial location.

Policies and achievements, 1953 to the late 1970s

Stalin's administration followed an extensive development policy, achieving greater output by increasing input. His successors realized that the approach was reaching its limits and wanted to switch to an intensive approach, which would produce more output per unit of input. This would require a more exact matching of technologies to tasks, and therefore the broadening of industrial priorities to include, for example, aluminium as well as steel, petrochemicals as well as coal-based chemicals, and electronic as well as mechanical engineering. Two obstacles were: firstly, that an isolated industrial economy, however big, could hardly expect to keep at the forefront of all technologies; and secondly, that the centralized control system built up hitherto was not well suited to managing a more sophisticated economy. The first obstacle was tackled with more determination and success than the second. Under Khrushchev, the Soviet government invested in many previously neglected industries. It also pursued economic integration with eastern European members of the CMEA, the organization established in 1949 to integrate communist-controlled economies. Some of these countries could supply technologies lacking in the USSR. Under Khrushchev's successors the USSR became far more active in world trade, selling raw materials and any manufactures already competitive on world markets in order to buy Western technology (see Ch. 9).

New national development policies brought new influences to bear on industrial location. At first sight, shifts in regional shares of output after 1953 seem very slight (Fig. 4.4). However, a 1 per cent spatial shift in production in the 1976–80 five-year plan would have involved about 70 times more production than in the 1928–32 plan. Shifts did take place and have continued to the present, reflecting firstly the new policies, secondly pressures on natural resources, transport and labour, and thirdly the outcome of debate on how to counter the pressures and implement the policies effectively.

Though east European industrial output and expertise were attractive to the Soviet government, eastern Europe lacked many of the natural resources needed to expand its industries and did not produce enough saleable goods to be able to buy what it needed on world markets. Integration with the Soviet economy therefore mainly involved the exchange of Soviet raw materials for east European manufactures. This, together with the expansion of Soviet domestic industries and the Soviet need to sell raw materials on world (hard-currency) markets if it wished to purchase Western technology, placed great strain on Soviet raw material production. From 1953 to the end of the Soviet period a large proportion of industrial investment necessarily went into extractive and primary processing industries. The biggest recipient was hydrocarbon production. Petroleum was the commodity most needed in eastern Europe, normally the one most saleable in Western markets, and, following the discovery of big oilfields well away from the frontiers, the preferred engine of Soviet industrial growth. Natural gas, though harder to market abroad, offered larger reserves and lower environmental damage in use. But pressures on supply affected other raw materials too, including coal, wood, most of the major metal ores, and even water.

Since raw materials had to be obtained, and maximum feasible national self-sufficiency remained a priority, there were few spatial alternatives for investment. Many of the shifts in industrial output after 1953 reflected the imperatives of resource location. The normal pattern was first to exploit resources closest to markets, such as the Volga–Urals oilfield, and then those further away, such as the West Siberian oilfields. Sometimes new technology made possible a return to resources closer to markets. This was so with the Kursk Magnetic Anomaly iron ores south of Moscow, which lay unattainably deep beneath waterlogged strata in the days before large-scale opencast mining. Their exploitation came after a long-term eastward shift of iron mining, which had almost reached Lake Baykal. But this was not the normal pattern. Rather, Soviet resource exploitation moved permanently further and further east, away from markets (Jensen *et al.* 1983). The over-used dictum, that 75 per cent of the population lived west of the Urals and 75 per cent of the natural resources lay to the east, added a new element, namely that the share of resource *extraction* east of the Urals began to approach that of resource *location*.

Since processing facilities and dependent industries, in any country, are usually set up either at the markets or at the first resources exploited, later-utilized resources are the most likely to be moved out with minimal processing. But in the Soviet case the distances between resources and markets became very great, and cheap sea transport was not generally available. Consequently there took place in the 1960s and 1970s a fierce debate on how to respond to the situation, recalling that between the proponents and opponents of Siberian industrial investment in the 1920s. As in the earlier debate, Siberianists advocated large-scale industrialization in their region, stressing plentiful natural resources, reduced strain on transport, and the official goal of spreading economic activity around the country. Europeanists pointed out that transport could be strained by inter-industry links too, and that severe natural conditions in Siberia meant high costs both for operating industry and transport and for enticing people to live there. The outcome was a compromise somewhat favouring the Europeanists. Huge quantities of raw materials moved westwards from Siberia, but processing and manufacturing expanded there too.

The concept of the Territorial Production Complex (TPC) was adopted in the 1970s to counter some of the attendant problems. TPCs (Fig. 4.6) are spatial concentrations of economic activity (except for one which is scarcely concentrated, the West Siberian oil and gas TPC), usually based on energy production and energy-intensive industries but including also ancillary manufacturing and servicing, and industries intended to round out employment opportunities. Thus the complex centred on the Bratsk hydro-electric station includes aluminium smelting, a pulp mill, and factories producing air-conditioning equipment and gloves. TPCs were intended to minimize intra-complex costs, focus inter-complex transport, and concentrate population for good access to employment opportunities and cultural, educational and medical facilities. However, high pay and generous benefits, introduced in the 1960s after the repudiation of Stalin's forced-labour policies, remained necessary to attract labour to Siberia.

Labour was hard to attract because opportunities for people with skills, and pleasanter living, were easy to find west of the Urals. Much of the country was short of labour because of the then current prodigal levels of manning. In Central Asia, however, there was still a large surplus rural population. This proved relatively immobile, reflecting perhaps poor education, including a poor command of Russian, and unwillingness to move to a very different cultural area. This situation stimulated another debate: should labour-intensive industries shift to Central Asia or remain where they were, mostly in European USSR, in the expectation that labour would eventually move to them? Once again the outcome seems to have favoured the European USSR. Although Central Asia's share of Soviet industrial output grew slightly after 1975, it even fell in the two decades after Stalin's death and remained at about half the region's share of the Soviet population.

Growing Soviet participation in trade outside the communist bloc revived the economies of the Black Sea, Baltic and Far Eastern seaboards. Ports were re-equipped and new ones added – Il'ichevsk and Yuzhnyy near Odessa, and Nakhodka and Vostochnyy east of Vladivostok, for example – and the Black Sea and Baltic ports added processing as well as transport facilities. Trade within the CMEA also favoured port or border locations for industry. The two sets of influences were reflected most clearly in the increasing shares of industrial production of the Baltic states and Belarus'.

The Urals and Volga regions present contrasting trends after 1953. The former remained dependent on the favoured heavy and defence-related industries of the Stalin era. Also, the exhaustion of some minerals, especially the iron ore of Magnitogorsk, forced a shift to lower-quality deposits and a reduction in the region's attractions for new investment. The neighbouring Volga region, however, steadily expanded its share of national industrial output, vindicating D.J.M. Hooson's assessment, a quarter-century ago, that it could become a 'new industrial heartland' (Hooson 1964). Its growth initially followed the opening-up of the Volga–Ural oilfield as the 'Second Baku', the successor to Trans-caucasia, in the 1950s and 1960s, but it then continued on the basis of an excellent location in the domestic economy. It lies between the old industrial areas, except the Urals, and the principal new sources of raw materials. It is less affected by pressure on local resources, such as water, than the former, but less remote from markets and linked industries than the latter. It has much better transport facilities than the Urals. It had recent experience in building and operating new industries, especially petrochemicals. Finally, a most important consideration for ministries facing production targets, its construction organizations had a good reputation, acquired partly through their work with Western firms which designed new factories in the region. This in turn helped them achieve priority in the allocation of personnel and materials. The region therefore acquired many industries in addition to the original petrochemicals, notably car and truck building (Fig. 4.7).

National security became a less prominent locational influence than it had been. After the appearance of inter-continental missiles, as well as the souring of Soviet–Chinese relations in the late 1950s, no part of the country was a particularly safe haven.

Soviet industry expanded rapidly after 1953, especially under Khrushchev, and it began to change in character. Diversification, changing attitudes to the outside world, and the choice of responses to changing natural resource and labour conditions constituted new influences on industrial location. Their impact might seem slight against the locational inertia of the bulk of industry (Shabad 1969). However, the amount of investment in new regions, though small as a percentage of the national total, looks large in relation to total national investment in industry in the years before, or even immediately after, the Second World War.

Industry from the late 1970s to the downfall of the USSR

From the late 1970s a new situation began to take shape as stagnation started to affect the economy as

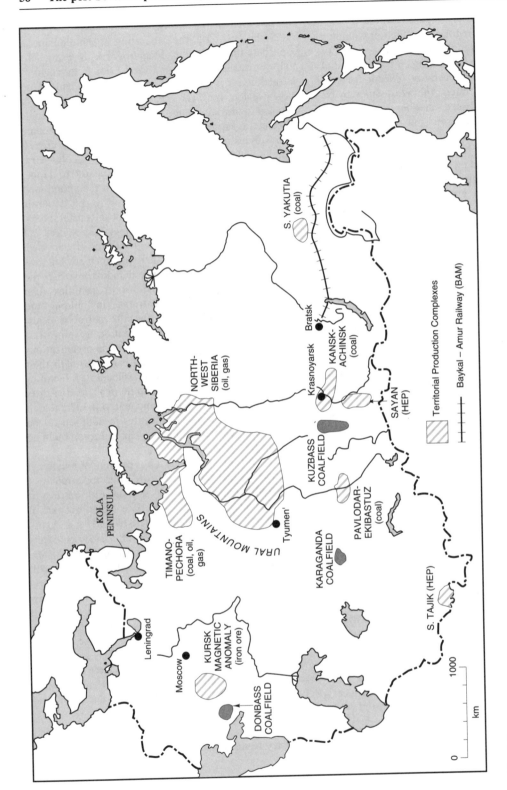

Fig. 4.6
Soviet territorial production complexes in the 1980s and associated energy and raw material sources.

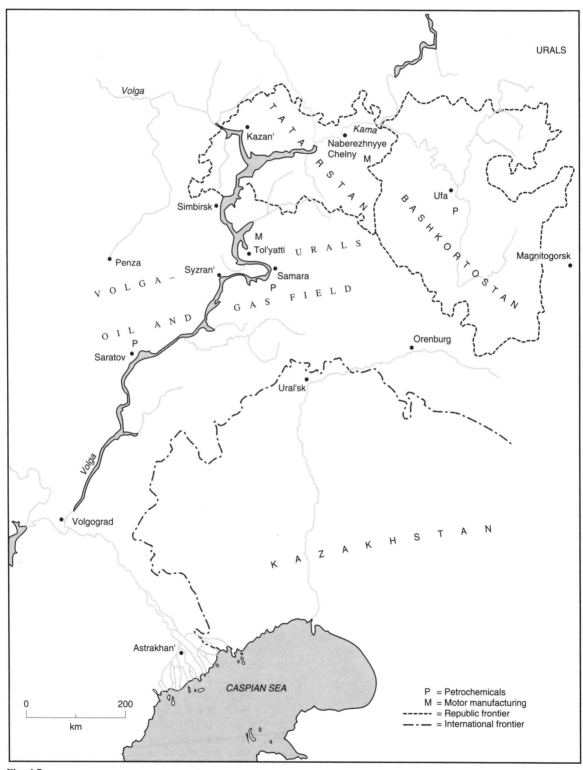

Fig. 4.7
The Volga economic region.

a whole including several key industries. Even steel, coal and petroleum production ceased to rise. There were two major responses to stagnation: the first an appeal to science and technology and the second an attempt to overhaul the economic management system. A further element in the new situation was developing environmental stress. This section will look briefly at the issues of science and technology and environmental stress before giving more extended consideration to economic management. Few people would have predicted at the beginning of the period that the attempt to confront economic stagnation would ultimately lead to the downfall of the USSR as a unified state.

Science and technology

Interest in foreign technology was already well established by the mid-1970s. The new phase was called 'The Scientific–Technical Revolution' in Brezhnev's time and was presented as a central part of *Perestroyka* under Gorbachev. It was focused less on buying new factories than on refurbishing what already existed, since much Soviet industrial plant was outdated and poorly maintained. Upgrading and renewing were to be a high priority, with increased investment planned to go into the retooling and modernization of enterprises, especially after 1985 under Gorbachev. A complementary emphasis was to be placed on building up research and putting its inventions into use more rapidly and efficiently. A long-standing problem of Soviet research was that many inventions, such as the continuous casting of steel, had been put into use abroad far more quickly than within the country. One of the intended benefits of these new policies was a reduction in industry's raw material needs, bringing in turn a reduction of pressure to invest in eastern resources.

Unfortunately, by the end of the 1980s, it had become clear that Gorbachev's investment policies were having the opposite effects to those intended. So ambitious were his investment targets that they had a destabilizing effect on the economy as a whole and contributed to the serious downturn in performance after 1989.

Environmental stress

Increasingly from the 1970s, stress on the natural environment became a source of serious concern. Environmental stress had many causes, but industrialization in the extensive mode characteristic of the centrally planned economy was clearly one. This was not merely a matter of growing shortages, such as water

shortages west of the Volga and in the south, or energy shortages across much of European USSR. A series of industrial disasters, most involving river pollution by chemical works in European USSR, helped change official attitudes to environmental misuse. Growing apprehension of air pollution was also a factor. The incidence of respiratory diseases reached alarming proportions in many industrial towns and even before the greater freedom of the Gorbachev years (the so-called era of *Glasnost'*) citizens' protests were frequently reported in the press. The 1980s, therefore, witnessed a tightening of legislation and an attempt to control pollution by levying penalties on offending enterprises. Under Gorbachev, the shock of the nuclear catastrophe at Chernobyl' helped stimulate an environmental movement and some enterprises were closed in consequence. Environmental concern also provided a vehicle for rising nationalism and thus contributed to the economic and political troubles of the later Gorbachev years.

Economic management

Criticism of the economic management system had surfaced from time to time after Stalin's death, particularly when performance fell below expectations. It was inevitable, therefore, that it would revive in the late 1970s with the onset of economic stagnation. The response was a rather timid attempt at economic reform in the early 1980s followed by more radical measures under Gorbachev. Before considering the latter and why they ultimately failed, it is worth examining the centrally planned system in detail and the problems to which it typically gave rise.

During Stalin's time economic decision making was highly centralized. Basic policy decisions, and decisions on some specific projects, were made at the top of the party and government hierarchies. Below that level, most economic power was concentrated in the all-Union ministries, especially those dealing with such priority industries as coal, steel and transport. Regional authorities had little economic power, as noted in Chapter 2. The big ministries tended to become semi-independent and mutually suspicious, mainly because of their constant battles for resources. They also initiated many of the proposals on which the higher authorities had to make policy decisions, and they could lobby powerfully when dealing with *Gosplan*, the agency responsible for controlling and co-ordinating their plans. All this meant that location decisions tended to reflect national and sectoral (ministerial) rather than regional or intersectoral viewpoints. Co-ordination of economic activities at the regional level was especially poor. Lines of communication to Moscow were long, and the co-ordination of local branches with those of another agency was usually a very low priority for any all-Union ministry. The Far Eastern economic region, for example, had to bring in most of its steel requirements along the Trans-Siberian railway, while the Komsomol'sk steelworks located in the Far East sent most of its output back along the same line, because it did not match local needs. Sometimes poor regional co-ordination reflected the lack of trust among all-Union ministries. For example, the railways ministry preferred to obtain coal from its own mines, even at the cost of very long hauls, rather than rely on the coal ministry.

Despite its deficiencies, the economic management system of Stalin's time served its purpose. The central government could mobilize resources rapidly and on a very large scale, and for extensive growth efficiency was adequate. But for the more intensive approach attempted after Stalin's death it was not adequate. Chapter 2 described how Khrushchev's administration in the 1950s abolished most of the all-Union ministries and transferred their power to regional authorities. But this simply replaced the tendency towards ministerial autarky by one towards regional autarky and raised costs. After the fall of Khrushchev, therefore, the country reverted to a system of all-Union ministries, albeit with more ministries than before in order to reduce the concentration of power. Intermittently, too, the government tried to enlarge the regional input into national planning. However, as long as the ministries were more powerful and were the chief initiators of planning proposals, regional inputs remained subordinate, a matter for the implementation rather than the initiation stage of planning.

The attempted shifts in control just described reflected two basic problems. The first was the inability of a highly centralized management to make and co-ordinate a large number and broad range of decisions efficiently. The second, as with any division of power, was the difficulty of preventing the pursuit ·of sub-national (sectoral or regional) interests when resources were scarce. Two further problems faced by the centrally planned economy concerned the range and quality of information available to and used by economic decision-makers. With respect to the first, during most of the Soviet era those agencies actually responsible for working out optimal factory locations concerned themselves primarily with production costs, and hardly at all with distribution costs, let alone total social and economic costs and benefits. The reason was that non-production costs were rarely their responsibility in what was, after all, a non-market economy. Total costs to the country as a whole were therefore often higher than they need have been. The

second and related problem was the fact that the quality of information available to decision-makers was frequently poor. The worst problem was the lack of realistic prices on which to base cost calculations, reflecting the fact that centrally-administered prices frequently bore little relationship to real costs. Use of such prices often imposed hidden costs on the economy. One example was the lavish use of natural resources which occurred especially under Stalin. This resulted from the underpricing of resources and from the use of below-cost transport tariffs for such goods. Use of such prices also encouraged the location of industries utilizing such resources in the old industrial regions west of the Urals.

One significant effect of the informational problems just described was the tendency to build enormous, highly specialized factories to supply the entire country while achieving large economies of scale. This tendency reached its apogee before the Second World War when it was denounced as 'gigantomania' but its effects are felt to this day. It resulted from excessive attention to on-site production costs. Over the years, costs and prices in the Soviet economy were gradually calculated with greater accuracy, but these refinements were no more able to avert stagnation in the 1970s than was the recourse to science and technology, discussed earlier.

The reforms of Gorbachev, known collectively as *Perestroyka*, began in 1985 with a rather conservative approach. This emphasized disciplinary measures to make the existing system work better. Campaigns against drunkenness, absenteeism and corruption were designed to raise labour productivity and improve morale. A new corps of state inspectors began to tighten quality control. At the same time small family enterprises were allowed to emerge, mainly in the service industries which the state had never operated efficiently. This represented a sharp change from earlier practice and its immediate effect was to begin to change the character of service and retail provision in the cities.

More radical measures followed from 1987. By this stage the Gorbachev leadership had evidently come to the conclusion that the economy could not be expected significantly to improve its performance within the parameters of the existing, centrally planned system. Therefore, it opted for decentralization and the injection of more marketization. This involved an attack on the central bureaucratic apparatus, including the industrial ministries which were regarded as the bulwarks of the old system. Ministries were now expected to confine their activities to long-term planning. The corollary to this was the granting of greater autonomy and financial responsibility to industrial enterprises. Particularly significant was the

Law on the State Enterprise, passed in 1987. According to this, over the next few years enterprises were to produce a declining proportion of their output in response to state orders (the old system) and an increasing proportion in response to orders derived from other enterprises in a market-type system. This was expected to produce greater efficiencies as firms were forced to meet their own costs and respond to market-type disciplines. Other reforms of the post-1987 period included more encouragement for private firms and particularly cooperatives (share ownership schemes were devised whereby state enterprises might have been converted to private ownership), allowing Soviet firms to participate directly in foreign trade, and permitting joint ventures to be established between Soviet and foreign firms for production within the Soviet Union.

Gorbachev's economic reforms failed to produce the expected results. An initial recovery over the period 1983–86 (possibly the result of the discipline campaign pursued after Brezhnev's death) was followed by economic downturn and rising inflation. The stagnation of the early 1980s was eventually succeeded by the economic dislocation of the end of that decade. Underlying reasons will be discussed below. But one very important factor seems to have been the failure by the leadership fully to appreciate the inbuilt inflexibility of the Soviet economic system. Attacking the planning bureaucracy was all very well, but the problem was that the functions it had performed could not be adequately performed by what was put in its place. The Law on the State Enterprise in particular had all kinds of undesirable side-effects including price increases, wage hikes, declines in output and a growing unreliability in the system of industrial supply. In the words of two commentators: 'although quite successful in dismantling the old system [Gorbachev] failed to create a viable new one' (Ellman and Kontorovich 1992:20).

Finally, then, why did Gorbachev's economic reforms fail? Some reasons have been suggested already: a destabilizing investment policy, and an inadequate programme for reforming the industrial management system. It is also the case that the Gorbachev administration made a number of mistakes in fiscal and monetary policy which helped fuel inflation. Finally, Gorbachev's democratic reforms have been cited as a cause. His various democratic innovations were designed to defeat the opponents of reform in party and government. Arguably, however, they worked in such a way as to reduce Gorbachev's chances of success by undermining the centre's legitimacy and giving free rein to the forces of public protest, separatism, criminality and dislocation.

The reasons for Gorbachev's failure are an issue about which future historians will debate endlessly. Even now there are different viewpoints. Some, for example, blame basic policy errors or the tactics of 'saboteurs', such as enemies of *Perestroyka* in the bureaucracy. Many suggest that sabotage could not have been defeated by any conceivable reform in view of the vested interests of planners, bureaucrats and other powerful groups germane to the central planning system. Yet others discount the importance of sabotage and argue that the mistake lay in trying to introduce market-type reforms into a centrally planned system in the first place. Two such scholars have suggested that a centrally planned system of the Soviet type requires continual downward pressure by the central authorities to survive (Ellman and Kontorovich 1992). Gorbachev, they maintain, reduced such pressure by weakening three essential supports of the system: the central bureaucratic apparatus, the official ideology, and the right of the party to intervene in the economy. Economic collapse, therefore, became inevitable. According to this view, maintaining the pressure and streamlining planning might have been enough to improve the system's performance in the medium term, though even then it might not have survived indefinitely.

After 1991: industry in the new republics

From what has been said above about Soviet industrial policy, it will be apparent that the newly independent republics are obliged to cope with the heritage left by a very particular type of development history. Most of the republics have inherited a heavy industrial sector developed for the specific purpose of serving Soviet goals, meaning an accent upon producer goods and manufacture for military purposes. The industrial profiles of the fifteen republics as they were in 1992 are illustrated in Table 4.1. The table indicates the significance to several of the republics of sectors such as fuel-energy industry, metallurgy, chemicals and petro-chemicals, and machine building and metalworking. The point is that, although the output of these sectors found a ready market in the old Soviet system, it is far less suited to a competitive world economy. Because of the inbuilt conservatism of the centrally planned economy, much of the heavy industrial plant is old and the products often lag behind the latest world standards. Lack of attention to costs in the past bodes ill for a market-oriented future. Moreover, such plant is often extremely resource- and labour-intensive which not only means higher costs but also, as noted above, unfortunate environmental side-effects.

Table 4.1 Share of industrial production by sector in the fifteen post-Soviet republics, 1992 (percentages of value of output)

	1	2	3	4	5	6	7	8
Armenia	11.7	3.6	7.8	27.2	0.9	2.9	12.9	14.4
Azerbaijan	32.2	7.7	6.2	14.0	0.8	1.9	16.8	17.0
Belarus'	15.9	2.1	11.7	27.9	4.6	4.6	13.3	13.1
Estonia	18.1	0.0	8.9	8.2	9.3	3.1	17.4	31.3
Georgia	4.3	4.6	4.5	8.7	3.6	4.0	20.4	37.3
Kazakhstan	12.4	14.8	4.6	7.7	1.8	4.9	18.5	25.0
Kyrgyzstan	4.5	3.2	0.7	15.9	1.8	4.1	37.4	26.5
Latvia	2.8	0.9	7.0	20.4	7.0	3.1	22.3	25.6
Lithuania	13.7	–	5.2	18.3	5.3	5.6	17.0	29.6
Moldova	10.0	–	0.8	14.4	3.8	4.2	12.1	48.3
Russia	27.1	18.3	9.2	21.1	4.6	2.5	7.4	9.8
Tajikistan	4.4	10.5	3.8	7.1	1.5	3.8	47.2	15.0
Turkmenistan	49.2	0.1	4.0	2.1	0.3	3.6	25.0	15.2
Ukraine	20.7	23.6	6.7	18.6	2.2	0.4	6.7	15.1
Uzbekistan	18.1	12.1	5.4	13.2	1.1	4.0	26.5	10.6

Source: *Statistical Handbook 1993: States of the former USSR (Studies of economies in transition, paper No. 8)* 1993 IBRD, World Bank, Washington DC: 18–19.

Key
1   Fuel-energy industry
2   Metallurgy
3   Chemicals and petrochemicals
4   Machinery building and metalworking
5   Forestry, woodworking, pulp and paper
6   Construction materials
7   Light industry
8   Agriculture/food processing.

There are other difficulties. As we have seen, under the Soviet system national production goals took priority over regional ones. The entire USSR was developed as a unified entity: a single 'USSR Ltd' (Nove 1977:30). Thus the imposition of a new set of international boundaries across what was previously a single economic space inevitably produces dislocation. This is especially the case when republics and even regions impose trade barriers to protect their economies from the difficulties being experienced by their neighbours, a process which began before the 1991 break-up. The effects are made worse by 'gigantomania'. Frequently there was monopoly or near-monopoly production of vital industrial goods by single enterprises which served the whole of the former USSR. With the complications introduced by the new international boundaries as well as by marketization and its many implications, firms often face difficulties in securing

their supplies or finding markets. Other features of the centrally planned economy have also left problems for the new republics. Industrial locations and transport networks which were created to serve the needs of the centrally planned economy within a single Soviet economic space are unlikely to be as well suited to the new situation.

Because of the accent placed upon national priorities and the problems which the lack of ministerial co-ordination produced for regional planning, the Soviet system gave rise to many regions with unbalanced or narrowly focused economies. This is an obvious difficulty for the new republics. The fact that Siberia and the Far East, for example, were turned by Soviet policy into energy and raw material suppliers for the European core, or that Central Asia became a 'plantation economy' (Dienes 1987) producing raw cotton for central Russian textile manufacturers, abetted numerous resentments. An insight into the overall economic structures of the republics is given by Table 4.2. Regional data indicating the bias of the Siberian and Far Eastern economies towards mining, fuel production and non-ferrous metals are provided in Table 4.3. Now that the USSR no longer exists, republics and regions are having to find new economic goals and address the imbalances bequeathed by Soviet planning.

Tables 4.4 to 4.6 give some indication of the economic turmoil which has attended the latter part of the Gorbachev era and the post-Soviet period. The reasons are many and frequently interlinked. Partly because of a loss of faith in central planning, and partly because of the actual breakdown of the command economy just before and then together with the disintegration of the USSR, the new republics have been trying to move along the path of marketization. But they have been doing so with varying degrees of enthusiasm and much uncertainty. One reason for this is the lack of agreement among republican leaderships and other powerful groups about either the need for such a change or how to go about it. Different individuals, groups and places are bound to be affected differently by such a momentous process, some benefiting and others suffering loss. In these circumstances, disagreement and political struggle are inevitable. A related reason has been the lack of a clear blueprint or established method for transforming a command economy into a market one. The processes occurring in eastern Europe and the former USSR are without historical precedent, and anyway each economy is unique, with its own peculiar set of circumstances. Mistakes have been made and further setbacks are probably unavoidable.

Disruption of linkages between different participants

Table 4.2 Sector shares of employment in the fifteen post-Soviet republics, 1991 (1985 in parentheses) (percentages)

	1	2	3	4
Armenia	23.3(20.03)	38.0(39.0)	10.7(12.0)	28.0(29.0)
Azerbaijan	33.7(33.2)	24.3(26.1)	12.8(13.6)	29.3(27.0)
Belarus'	19.1(23.6)	41.5(38.8)	13.6(14.3)	25.8(23.2)
Estonia	12.4(13.3)	43.8(42.4)	17.5(17.1)	26.2(27.3)
Georgia	26.9(27.9)	28.7(28.7)	13.2(13.6)	31.3(29.8)
Kazakhstan	26.1(23.7)	34.3(31.7)	18.6(18.5)	20.9(26.0)
Kyrgyzstan	35.5(32.8)	26.5(27.3)	11.0(13.0)	27.0(26.9)
Latvia	16.3(15.1)	40.3(40.9)	16.8(17.7)	26.5(26.3)
Lithuania	17.8(19.7)	39.5(40.4)	15.7(14.7)	27.0(25.3)
Moldova	35.9(36.4)	27.9(27.8)	11.2(13.1)	25.1(22.6)
Russia	13.5(14.3)	41.9(41.7)	14.6(16.3)	30.1(27.7)
Tajikistan	44.8 (43.2)	20.5(21.4)	10.4(11.6)	24.3(23.8)
Turkmenistan	42.4(40.4)	20.8(20.9)	11.4(14.0)	25.3(24.7)
Ukraine	19.3(21.2)	40.2(39.1)	13.4(14.5)	27.1(25.2)
Uzbekistan	41.9(37.9)	22.5(23.1)	10.5(12.6)	25.2(26.3)

Source: *Statistical Handbook 1993: States of the former USSR (Studies of economies in transition, paper No. 8)* 1993 IBRD, World Bank, Washington DC: 26–701.

Key
1. Agriculture and forestry
2   Industry and construction
3   Other material, e.g. transport of goods, road maintenance, wholesale trade
4   Non-material, e.g. general transport, housing, public services.

in the economy has already been mentioned as a factor inhibiting marketization. This has many causes, but one significant one has been the accumulation of debts where firms have been unable to secure payment for their goods (perhaps because their customers have

Table 4.3 West and East Siberia's contribution to Russian resource production, 1980–90 (as a percentage of Russia's total)

	1980	1985	1987	1990
Oil	57.2	67.9	71.9	72.7
Gas	63.0	82.4	85.4	88.8
Coal	57.4	59.4	61.0	61.3*
Wood and timber	28.7	30.5	31.5	31.3*
Electricity	27.9	26.0	26.7	27.8*

Source: Bradshaw M J 1992, Siberia poses a challenge to Russian federalism. *RFE/RL Research Report* **1**(41), 16 October: 6–14.
* 1989 data

Table 4.4 The post-Soviet republics: average annual growth of net material product (NMP), 1985–92 (per cent)

	1985–90	1991	1992
Armenia	0.1	−11.6	−46.0
Azerbaijan	−2.2	−1.9	−31.2
Belarus'	2.9	−1.9	−10.6
Estonia	–	–	–
Georgia	−3.2	−20.6	−39.7
Kazakhstan	–	−14.9	−14.3
Kyrgyzstan	4.8	−4.3	–
Latvia	3.6	−3.8	–
Lithuania	4.4	−9.3	–
Moldova	3.6	−18.0	−21.3
Russia	1.7	−14.3	–
Tajikistan	1.8	–	–
Turkmenistan	2.6	−4.7	–
Ukraine	0.5	−13.4	–
Uzbekistan	4.7	−3.7	−14.4

Source: *Statistical Handbook, 1993: States of the former USSR (Studies of economies in transition, Paper No. 8)* 1993 IBRD, World Bank, Washington DC: 8–9.

Table 4.5 Commonwealth of Independent States: oil production by member states, 1985–92 (millions of tonnes)

	1985	1986	1987	1988	1989	1990	1991	1992
Azerbaijan	13.2	13.3	13.8	13.7	13.2	12.5	11.7	11.0
Belarus'	2.0	2.0	2.0	2.1	2.1	2.1	2.1	2.0
Kazakhstan	22.8	23.7	24.5	25.5	25.4	25.8	26.6	27.5
Kyrgyzstan	0.2	0.2	0.2	0.2	0.2	0.2	0.1	0.1
Russia	542	561	569	569	552	516	461	396
Tajikistan	0.4	0.4	0.3	0.3	0.2	0.1	0.1	0.1
Turkmenistan	6.0	6.0	5.8	5.7	5.8	5.7	5.4	5.3
Uzbekistan	2.0	2.2	2.3	2.4	2.7	2.8	2.8	3.1
Ukraine	5.9	5.7	5.6	5.4	5.4	5.3	4.9	4.4
CIS total*	595	615	624	624	607	570	515	449
Georgia	0.6	0.2	0.2	0.2	0.2	0.2	0.2	0.1

Source: *Mir v tsifrakh. Statisticheskiy sbornik, 1992* Statisticheskiy komitet Sodruzhestva nezavisimykh gosudarstv. Finansovyy inzhiniring, Moscow: 170–1; Sagers M J 1993 The energy industries of the former USSR: a mid-year survey. *Post-Soviet Geography* **34**(6), June: 341–418.

* excluding Georgia

Table 4.6 Commonwealth of Independent States: coal production by member states, 1985–92 (millions of tonnes)

	1985	1986	1987	1988	1989	1990	1991	1992
Kazakhstan	131	138	142	143	138	131	130	127
Kyrgyzstan	3.9	4.0	3.9	4.0	4.0	3.7	3.5	2.2
Russia	395	408	415	425	410	395	353	337
Tajikistan	0.5	0.7	0.6	0.7	0.5	0.5	0.3	0.2
Uzbekistan	5.3	6.0	5.0	5.5	6.2	6.5	5.9	4.7
Ukraine	189	193	192	192	180	165	136	134
CIS total*	725	749	758	770	739	702	629	605
Georgia	1.7	1.7	1.6	1.4	1.2	1.0	0.7	0.5

Source: *Mir v tsifrakh. Statisticheskiy sbornik, 1992* Statisticheskiy komitet Sodruzhestva nezavisimykh gosudarstv. Finansovyy inzhiniring, Moscow: 182–183; Sagers M J 1993 The energy industries of the former USSR: a mid-year survey. *Post-Soviet Geography* **34**(6), June: 341–418.

* excluding Georgia

also fallen victim to bad debts or are suffering from the effects of inflation) and thus cannot pay their own suppliers and workers. Indebtedness has been encouraged by the loose budgeting which was so characteristic of the command economy and which has continued into the new era. Multiplied many times over, mounting debts ultimately undermine the economy as a whole. Payments difficulties are exacerbated where customer and supplier find themselves on either side of the new frontiers. Linkages can then be disrupted by political factors, as noted already, or by economic ones such as the complications

brought about by the collapse of the ruble zone. Several republics have been introducing their own new currencies or substitute ones, but even if these were stable the net effect would be to increase the costs of trade. The establishment of the Commonwealth of Independent States (CIS) between eleven (now twelve) of the post-Soviet republics at the end of 1991 was supposed to minimize such difficulties by guaranteeing the continued existence of a single economic space and other forms of inter-republican cooperation. In practice, however, such cooperation has been far from easy to achieve.

Among other causes of economic dislocation have been the collapse of the CMEA at the end of the 1980s which undermined many long-established industrial linkages, and the endemic criminality and disorder. Armed conflict has had a detrimental effect, particularly in the south.

The general economic collapse has encouraged the widespread growth of 'primitive' activity such as bartering, as well as the flourishing of the household economy in such areas as food production. Ironically enough, there is plenty of precedent for such 'informal' activity in the command economy, but it has probably never been as important as it is now. Whether this will be a permanent feature of the new age remains to be seen.

At the time of writing, it is Russia and the Baltic states which have moved furthest along the path of transforming themselves into market economies, but even there much remains to be done and eventual success is by no means assured. The policies pursued include decontrolling prices, giving enterprises autonomy, removing subsidies, aiming at macroeconomic stability (controlling inflation, balancing national budgets), providing a safety net for those people worst affected by the reforms, and opening up the economy to foreign trade and investment. The ultimate aim is to encourage efficiency by allowing uncompetitive industries to fail and the successful to flourish without interference by the state. Privatization is also seen as essential in view of the obvious failures of state ownership in the communist era. But privatizing is far from easy in societies without an established capitalist class and where much of the industry is uncompetitive and thus unattractive to both domestic and foreign investors. It therefore takes time. Other reforms now on the agenda include monetary reform, making currencies convertible, and providing the necessary infrastructure of financial, legal and management functions to enable a capitalist economy to work.

Ukraine, by contrast, has been one of the republics which are more hesitant about reform. Here, although rather similar policies of marketization and privatization have been introduced, the government has been slow to implement them and continues to subsidize old-established industries to a marked degree. Ukraine's problems have been complicated by its reliance on traditional heavy industry and its need to import fuel.

The degree to which reform policies are pursued and are successful depends upon a multiplicity of factors, including the internal political struggle and external pressures (for example, those exerted by international agencies like the IMF). A key question concerns the fate of the military–industrial complex, that interlinked system of state-owned enterprises with military connections which was central to the command economy and which still forms a significant element in the post-Soviet economies. Can the republics afford to cease subsidizing such activities and thereby foster greatly increased unemployment and the hostility of powerful vested interests? Equally, how long can they afford to continue subsidies with the certainty that this will delay economic reform? Present policies of converting many military industries to civilian production may be only partially successful. An alternative is to allow such industries to export their products, especially the high technology ones, even though the world market for arms is extremely competitive.

Just as world capitalism is currently undergoing considerable restructuring with varied effects nationally, regionally and locally, so the same must now follow in the post-Soviet economies. What cannot be predicted with any certainty is what the repercussions will be for the different republics and regions. So much depends upon the international economy, government policy (including regional policy) and local responses. Will marketization, for example, lead to the setting up of new industries and services with a modern orientation, and if so where – in capital cities, regional and communications centres, at ports, close to frontiers and major routeways? Will a region like Siberia with rich natural resources receive the capital necessary to sell those resources on the world market, and will Siberians themselves benefit? Siberians will no doubt wish to weigh the financial advantages to be gained from resource exploitation against the potential disadvantages of resource depletion, environmental disruption and possible over-dependence on foreign investors. Other questions also come to mind. What, for example, will happen to the big state enterprises, particularly those which formed part of the military–industrial complex, and to those regions of heavy industry formed during the industrialization drive of the Stalin years? Will activities which were developed to serve the autarkic policies of the past and badly located with respect to the needs of the present now disappear? What about the many regions without rich resources, and relatively backward ones like Central Asia with unbalanced economies? Within the Russian Federation, future industrial location patterns are likely to be influenced by the balance of political power between Moscow and the regions. If Moscow retains the upper hand, the future may see a continuation of the core–periphery relationship of the past, while the devolution of economic and political power to the regions may encourage the evolution of more balanced economies in parts of the periphery.

While unemployment across the former USSR has

so far been selective (hitting, for example, the central Russian textile industry cut off from its Central Asian suppliers), the subsidies which have permitted it from spreading cannot continue indefinitely. Recent evidence from Russia suggests that heavy industrial regions may suffer disproportionately from the effects of restructuring, although some military-type activities with high technology may fare better. Towns with single industries or only one or two activities, of which there are many across the former USSR, will certainly be vulnerable. The possibility that mass unemployment may lead to sustained political or even violent opposition to reform rightly worries the marketizers.

Across the former USSR the supporters of market-type reform contend that the present economic problems are only temporary and will eventually give way to stable societies characterized by healthy economic growth. The opponents are much less optimistic, but few are able to offer coherent alternative policies. What is certain is that radical economic restructuring is now underway and that the emerging industrial geography, though no doubt still owing something to the past, will nevertheless be very different from that inherited from the years of Soviet industrialization.

References

Dienes L 1987 *Soviet Asia: Economic development and national policy choices.* Westview, Boulder
Ellman M, Kontorovich V 1992 *The Disintegration of the Soviet Economic System.* Routledge, London
Hooson D J M 1964 *A New Soviet Heartland?* Van Nostrand, Princeton
Jensen R G, Shabad T, Wright A W (eds) 1983 *Soviet Natural Resources in the World Economy.* University of Chicago Press, Chicago
Khrushchev A T, Nikol'skiy I V 1985 *Ekonomicheskaya geografiya SSSR, chast' 1,* 2nd edn. MGU, Moscow
Nove A 1977 *The Soviet economic system.* Allen and Unwin, London
Shabad T 1969 *Basic Industrial Resources of the USSR.* Columbia University Press, New York

5

Transport

Robert N. North

Introduction

Transport networks and traffic flows reflect the interplay of many factors, but some of them are particularly important in the former Soviet Union. Firstly, there are physical-geographic realities which cannot easily be modified by human action: climate, topography and the locations of natural resources. Secondly, there are national policies for economic development in general and transport in particular. And, thirdly, there are man-made aids to and constraints on policy implementation, some deriving from technology and some from the nature of the political–economic system. This chapter surveys the major factors. In the case of policies and constraints it takes an historical approach, making divisions at 1953 and the end of 1991, on the grounds that many policies and decisions of past times have left an enduring legacy. Finally, some possibilities for the future are examined.

Climate and topography

The former Soviet Union is well supplied with long navigable rivers, most of them in Russia (Fig. 5.1). In past centuries they were the principal long-distance transport mode, but they have not retained their pre-eminence. In a modern industrial economy freight must be able to move year-round, and the inland waterways cannot offer year-round service. In most of the country they freeze for several months (Fig. 5.1). In the hot, dry regions of southern Siberia, Kazakhstan and Central Asia they also suffer from low water in summer, a problem exacerbated by the competing demands of irrigation and power generation. Eastern Siberia and the northern Far East are also affected by summer low water, a consequence in their case of hilly terrain, permafrost and low precipitation. Surface moisture cannot seep into frozen ground, so rivers in the region typically have a rapid spring runoff after snowmelt, followed by extremely low water during the summer. Winter freezing and summer low water together leave many rivers with navigation seasons of less than a month, and some with as little as ten days.

Rivers in the former Soviet Union are often said to run in directions which do not match major traffic flows, but this is only partly true. Certainly most rivers run north–south or south–north, while the biggest traffic flows run east–west. But there are many navigable east–west tributaries, and those in adjacent river basins are often close enough to permit easy canal connections. Furthermore, even north–south traffic flows are in the tens of millions of tonnes both east and west of the Urals. Almost without exception, length of navigation season is a far greater problem than location or direction.

Unlike the inland waterways, most maritime transport can operate through the winter. Only the Arctic coast east of the River Yenisey now closes down completely, but to keep sea lanes open elsewhere the former USSR maintained the world's largest fleet of icebreakers and icebreaking freighters. This costly undertaking has now fallen mainly on Russia, though even the northern parts of the Black and Caspian Seas are affected by ice.

The interrupted nature of the coastline severely restricted the utility of sea transport for the Soviet Union. Most domestic maritime traffic stayed within individual bordering seas: the Black Sea, the Baltic, the western Arctic, and the Sea of Okhotsk and neighbouring waters. The maritime routes linking them were too roundabout to be worth using on any significant scale. In the mid-1980s only 1 per cent of Far Eastern maritime traffic was with other parts of the country.

Road transport under the Soviet regime was less well developed than in almost any other industrial country, and the discouraging natural conditions were partly to blame. Winter problems include snow blowing, frost damage and, in permafrost regions, buckling of the whole road bed. In the extreme cold of the north-east, vehicles become inoperable unless specially adapted. During the rapid spring thaw, characteristic of continental climates, poor-quality roads may be impassable for weeks. In the oil and gas regions of north-western Siberia there are huge areas of swamp, where the building of all-weather roads is prohibitively expensive. And parts of Kazakhstan and Central Asia are affected by blowing sand. On top of all this, the average lengths of haul for most commodities are much greater than the distances at which road transport can compete with railways (Tables 5.1 and 5.2). It is true that there are large dry areas in Kazakhstan and Central Asia where motor transport requires little or no road building. Also winter (ice) roads in the north, though expensive to operate, are much cheaper than rail or air transport for small traffic volumes and can offset the seasonal loss of river transport. Furthermore, much of European Russia is no worse climatically than parts of North America which have heavy road traffic. Nevertheless, natural conditions have helped to ensure that the great bulk of road traffic moves locally over very short distances.

Of all the surface modes of transport, railways and pipelines have proved best able to cope with the natural conditions. They too can suffer permafrost damage unless expensively constructed and are costly to build

Table 5.1 Average hauls, common-carrier transport (kilometres)

	1913	1953	1985	1988	1990
Railway	484.8	747.6	941.1	953.5	960.0
Sea	1 344.4	1 059.3	3 775.6	3 935.4	4 125.3
Waterway	823.4	508.6	413.4	363.5	347.5
Pipeline	750.0	258.5	2 080.7	2 351.0	2 365.8
Motor	10.0[1]	10.5[1]	18.4[1,2]	20.7	21.4
Air	–	1 000.0	1 046.9	1 030.3	1 103.4
Total	577.5[1]	221.7[1]	212.9[1]	661.7	642.6

Sources: Tables 5.3 and 5.4

Notes: 1. Includes all motor transport
2. Common carrier only: 22.4

Table 5.2 Railway freight dispatched, common carrier, by commodity, USSR, 1985 and 1990 (mill.tonnes, thou.mill.tonne-km, km)

	Weight	Per cent	Volume	Per cent	Av.haul
1985					
Coal and coke	791.0	20.0	640.5	17.2	809.7
Oil and products	420.6	10.6	436.8	11.7	1 038.5
Iron and steel	205.3	5.2	310.8	8.5	1 513.9
Forest products	151.6	3.8	259.2	7.0	1 709.8
Grain, milled products	147.2	3.7	157.5	4.2	1 070.0
Ore	326.1	8.3	254.8	6.9	781.4
Mineral building materials	1 037.4	26.3	521.4	14.0	502.6
Fertilizers	146.7	3.7	159.1	4.3	1 084.5
Other	725.3	18.4	978.3	26.3	1 348.8
Total	3 951.2	100.0	3 718.4	100.0	941.1
1990					
Coal and coke	775.6	20.0	685.6	18.4	884.0
Oil and products	393.0	10.2	430.6	11.6	1 095.7
Iron and steel	194.2	5.0	294.0	7.9	1 513.9
Forest products	142.9	3.7	262.3	7.1	1 835.5
Grain, milled products	150.6	3.9	185.4	5.0	1 231.1
Ore	318.2	8.2	242.6	6.5	762.4
Fertilizers	138.3	3.6	162.4	4.4	1 174.2
Other	1 759.2	45.4	1 454.2	39.1	826.6
Total	3 872.0	100.0	3 717.1	100.0	960.0

Sources: TsSU SSSR, *Narodnoye khozyaystvo SSSR v 1985 g. Statisticheskiy yezhegodnik* (Moscow: Statistika, 1986), pp. 326, 327; Informatsionno-izdatel'skiy tsentr Goskomstata SSSR, *Transport i svyaz'. Statisticheskiy sbornik* (Moscow, 1991), pp. 42–4.

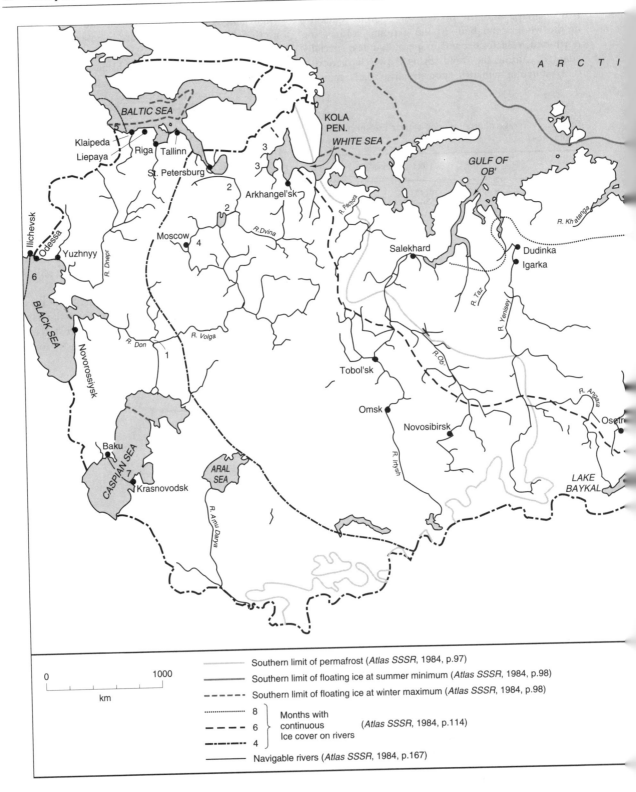

ARCTI

BALTIC SEA

KOLA
PEN.

WHITE SEA

Klaipeda
Liepaya
Riga Tallinn
St. Petersburg

3
3
2
2

Arkhangel'sk

GULF OF
OB'

R. Pechora

R. Khatanga

Ilichevsk
Odessa
Yuzhnyy

Moscow 4

R. Dvina

Salekhard

Dudinka
Igarka

6

R. Dnepr

BLACK SEA

Novorossiysk

R. Don

R. Volga

1

R. Taz

R. Yenisey

Tobol'sk

R. Ob'

R. Angara

Omsk

Novosibirsk

Osetr

Baku

CASPIAN SEA

1

Krasnovodsk

ARAL
SEA

R. Irtysh

LAKE
BAYKAL

R. Amu Darya

0	1000
	km

Southern limit of permafrost (*Atlas SSSR*, 1984, p.97)

Southern limit of floating ice at summer minimum (*Atlas SSSR*, 1984, p.98)

Southern limit of floating ice at winter maximum (*Atlas SSSR*, 1984, p.98)

8 Months with
6 continuous   (*Atlas SSSR*, 1984, p.114)
   Ice cover on rivers
4

Navigable rivers (*Atlas SSSR*, 1984, p.167)

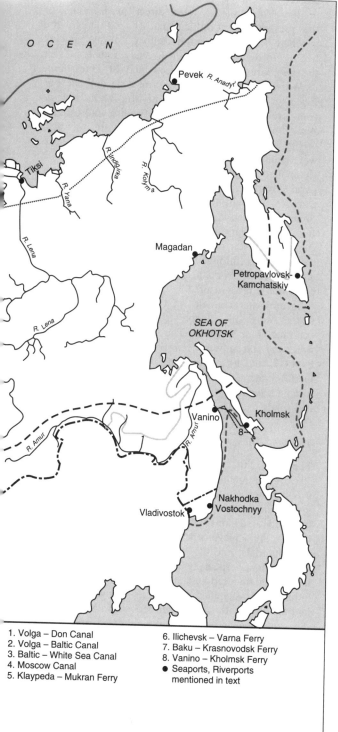

1. Volga – Don Canal
2. Volga – Baltic Canal
3. Baltic – White Sea Canal
4. Moscow Canal
5. Klaypeda – Mukran Ferry

6. Ilichevsk – Varna Ferry
7. Baku – Krasnovodsk Ferry
8. Vanino – Kholmsk Ferry
● Seaports, Riverports
   mentioned in text

Fig. 5.1
Water transport in the former USSR.

through swamps. But high construction costs in a few regions are offset by low costs elsewhere, by the ability of both modes to carry heavy traffic in all seasons, and by their suitability for long-distance traffic. They are also favoured by topography. With exceptions in the Caucasus and east of the River Yenisey, railways in the former USSR have milder gradients and fewer tunnels than any other large national system.

The relative distribution of natural resources and markets

Soviet writers made much of the growing spatial separation between population ('three-quarters west of the Urals') and natural resources ('three-quarters east of the Urals'). The resources referred to are principally minerals, forests and hydro-electricity: the better agricultural lands lie mainly in Europe. Indeed good agricultural land formed the basis of nineteenth-century population distribution, since 80 per cent of the population was rural, and by extension the basis for population distribution today. When non-agricultural resources were needed, those in the populated regions were exploited first. Then the search moved outwards and primarily eastwards, except when technological change – in mining or prospecting, for example – shifted attention back to the populated regions. Resource depletion in the west and the addition of markets in eastern and western Europe have brought about the present extreme spatial imbalance, but the basic situation is by no means new. Furs from the north in the sixteenth and seventeenth centuries, gold and copper from the Altay mountains in the eighteenth century, and wheat and dairy produce from the West Siberian plain in the nineteenth and early twentieth centuries all faced essentially the same problem, namely the sheer inaccessibility of Siberia, coupled with its inability to attract enough settlers to form a substantial market. The search for efficient long-distance freight transport is a persistent theme in Russian economic history, and the current reliance on railways and pipelines represents only the latest stage in a continuing saga.

Policies and technologies

The Soviet inheritance from tsarist Russia

Until the mid-nineteenth century, long-distance transport in the Russian Empire depended on rivers, joined by canals in European Russia, and a system of post-roads. Both the canals and the post-roads stemmed from initiatives of Peter the Great in the early eighteenth century. The canals, linking St Petersburg to the Volga basin, were direct predecessors of the present Volga–Baltic canal (Fig. 5.1). In the last half of the nineteenth century railways quickly became pre-eminent. By 1914 they radiated from the two largest cities, Moscow and St Petersburg, to Warsaw, the Caucasus, Central Asia and the Far East, serving military and administrative as well as commercial functions (Fig. 5.2). The line into Central Asia from Krasnovodsk, the Caspian port, was also originally a military line. Other parts of the network were primarily commercial. Lines in western Ukraine channelled wheat for export to Odessa, and a set of parallel routes linked the coal of the Donbass with the iron ore of Krivoy Rog. The biggest pre-revolutionary project, the Trans-Siberian railway, filled several roles. It carried settlers and troops to Siberia and the Far East, served the export of West Siberian wheat and dairy produce through the Baltic ports, and provided a transit route from the Far East to western Europe for Chinese tea and silk.

Several characteristics of the late tsarist transport system were still clearly visible in 1953, and some continue to be so today. The skeletons of the present waterway and railway networks were already in place by 1917, and little was done during Stalin's time to modify that aspect of transport which most struck foreign travellers before the revolution: the appallingly bad roads.

The legacy of the Stalin era

There was little new transport construction from the 1917 revolution to the late 1920s, but the Stalin era was another matter, despite a policy of minimizing investment in transport. National development policies included rapid industrialization and its attendant urbanization, the achievement of self-sufficiency in strategic raw materials, and the territorial spread of economic development, especially to the eastern regions. The last-named focused on mining and associated heavy industry. It was typified by the Urals–Kuznetsk Combine and the exploitation of copper deposits in Kazakhstan. An additional policy, partly deliberate and partly forced by the collapse of world markets after 1929, was to switch from the tsarist reliance on commodity exports to a much more internalized economy. All of these policies placed heavy demands on transport: for routes to natural resources; for better rural–urban transport to supply the new and larger towns; and for new inter-regional links and the upgrading of existing ones to carry heavy industrial traffic.

The desire to minimize investment in transport derived from the fact that it does not produce material goods. 'Place utility' was not prominent in traditional Marxist economic reasoning. Minimization of investment during Stalin's time took the form of, firstly, choosing one general-purpose mode, the railways, and concentrating investment therein; secondly, establishing a few major railway axes, concentrating as much traffic as possible on them, and thereby achieving economies of scale; and thirdly, relying heavily on organizational refinements to increase capacity. The axes were three in number (Fig. 5.2): the Trans-Siberian from Moscow to Irkutsk; Leningrad–Moscow–Donbass–Rostov–Caucasus; and Donbass–Krivoy Rog. 'Axis' did not necessarily mean a single route: between Moscow and the Donbass three were upgraded to provide the required capacity. Apart from the axes, most new railway lines of the Stalin era were either inter-regional links, such as that from the Urals to the Donbass, or lines to mineral resources.

A result of the Stalin-era approach to transport was that by 1953, despite heavy destruction in the Second World War, the Soviet Union could boast a railway system comparable with any in the world in terms of traffic moved and organizational efficiency. The focusing of investment in the railways was strongly reflected in traffic statistics. National freight traffic had grown 7.5 times since 1913, in tonne-kilometres, and the railways' share had risen from 61 to 85 per cent (Table 5.3). Waterway and sea traffic had only

doubled, and while motor transport accounted for 71 per cent of tonnes loaded, the average haul was only 10 kilometres (Tables 5.1, 5.4). That is not to say that the transport legacy of the Stalin era consisted only of new and rebuilt railways. The White Sea–Baltic canal was opened in 1933 and the Volga–Don canal in 1952 (Fig. 5.1). An internal air service was created, and the Moscow metro (underground railway) was a national showpiece. But roads were primitive by Western standards, and the supply of motor vehicles was pathetically small, so that remoteness from a railway could mean virtual inaccessibility. Seaports were poorly equipped, and cargo vessels, used largely for coastal traffic, averaged about one-third the size of their Western counterparts.

Transport in the post-Stalin USSR

National development policies and their consequences for transport. Some policies of Stalin's time persisted after his death. The USSR continued to rely mainly on domestic natural resources, though the policy was relaxed for grains and some minerals, notably bauxite. Other policies changed. Of particular importance to transport were the re-emergence of foreign trade, a move away from coal as the dominant energy source, and an expansion of military activity.

Table 5.3 Traffic volumes, common-carrier transport, USSR (thou.mill.tonne-km and percentage shares)

	1913[1]		1953		1985		1988[2]		1990	
	Volume	Per cent	Volume	Per cent	Volume	Per cent	Volume	Per Cent	Volume	Per Cent
Railway	76.4	60.6	798.0	84.5	3 718.4	47.6	3 924.8	45.6	3 717.1	44.7
Sea	20.3	16.1	48.2	5.1	905.0	11.6	1 011.4	11.7	944.7	11.4
Waterway	28.9	22.9	59.3	6.3	261.5	3.3	251.2	2.9	232.5	2.8
Pipeline	0.3	0.2	7.6	0.8	2 443.1[3]	31.3	2 917.6	33.9	2 898.1	34.8
Motor[4]	0.1	0.1	31.4	3.3	476.2[5]	6.1	508.0[6]	5.9	526.8[7]	6.3
Air	–	–	0.2	–	3.4	–	3.4	–	3.2	–
Total	126.0	100.0	994.7	100.0	7 807.6[8]	100.0	8 616.4	100.0	8 322.4	100.0

Sources: TsSU SSSR. *Transport i svyaz' SSSR, Statisticheskiy sbornik* (Moscow: Statistika, 1972), p. 17; TsSU SSSR, *Narodnoye khozyaystvo SSSR v 1985 g.. Statisticheskiy yezhegodnik* (Moscow: Statistika, 1986), p. 323; Informatsionno-izdatel'skiy tsentr Goskomstata SSSR, *Transport i svyaz', Statisticheskiy sbornik.* (Moscow, 1991), pp. 42, 46; Goskomstat SSSR, *Transport i svyaz' SSSR. Statisticheskiy sbornik* (Moscow: Finansy i statistika, 1990), pp. 30, 38

Notes: 1. Boundaries of post-Second World War USSR
2. The peak year for traffic
3. Oil 1 312.5, gas 1 130.6
4. Includes non-common carriers
5. Common-carrier 141.6
6. Common-carrier 143.3
7. Common-carrier 135.7
8. Common plus non-common carrier totalled 8 772.8.

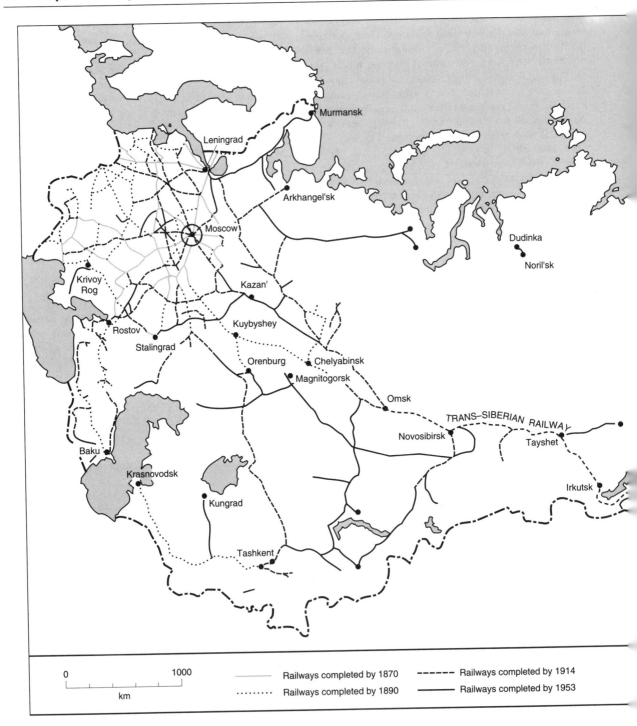

Murmansk

Leningrad

Arkhangel'sk

Moscow

Dudinka

Noril'sk

Krivoy
Rog

Kazan'

Kuybyshey

Rostov

Stalingrad

Orenburg   Chelyabinsk

Magnitogorsk

Omsk

TRANS–SIBERIAN RAILWAY

Novosibirsk

Tayshet

Baku

Krasnovodsk

Irkutsk

Kungrad

Tashkent

0 1000	
km	

——— Railways completed by 1870   --- Railways completed by 1914

········· Railways completed by 1890   ——— Railways completed by 1953

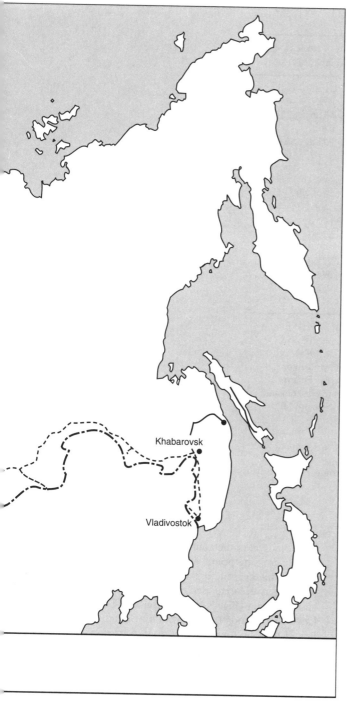

Fig. 5.2
Railway development to 1953

Table 5.4 Tonnes dispatched, common-carrier transport, USSR (millions and percentage shares)

	1913[1]		1953		1985[2]	
	Weight	Per cent	Weight	Per cent	Weight	Per cent
Railway	157.6	72.2	1 067.4	25.0	3 951.2	12.6
Sea	15.1	6.9	45.5	1.1	239.7	0.8
Waterway	35.1	16.1	116.6	2.7	632.6	2.0
Pipeline	0.4	0.2	29.4	0.7	630.8	2.0
Motor	10.0	4.6	3 002.7	70.5	25 900.0	82.6
Air	–	–	0.2	–	3.2	0.01
Total	218.2	100.0	4 261.8	100.0	31 357.5	100.0

	1985[3]		1988		1990	
	Weight	Per Cent	Weight	Per Cent	Weight	Per Cent
Railway	3 951.2	32.2	4 116.0	31.1	3 872.0	31.4
Sea	239.7	2.0	257.0	2.0	229.0	1.9
Waterway	632.6	5.2	691.0	5.2	669.0	5.4
Pipeline	1 113.0	9.1	1 241.0	9.4	1 225.0	9.9
Motor	6 320.0	51.5	6 921.0	52.3	6 344.0	51.4
Air	3.2	0.01	3.3	0.01	2.9	0.02
Total	12 259.7	100.01	13 229.3	100.01	12 341.9	100.02

Sources: TsSU SSSR. *Transport i svyaz' SSSR. Statisticheskiy sbornik* (Moscow: Statistika, 1972), p. 21; TsSU SSSR, *Narodnoye khozyaystvo SSSR v 1985 g.. Statisticheskiy yezhegodnik* (Moscow: Statistika, 1986), pp. 326, 330, 335, 336, 350; Gosudarstvennyy komitet SSSR po statistike, *Transport i svyaz' SSSR. Statisticheskiy sbornik* (Moscow: Finansy i statistika, 1990), p. 8; Informatsionno-izdatel'skiy tsentr Goskomstata SSSR, *Transport i svyaz'. Statisticheskiy sbornik* (Moscow, 1991).

Notes: 1. Boundaries of post-Second World War USSR.
2. Motor transport includes non-common carrier traffic. Compatible with preceding years.
3. Motor transport includes only deliveries by common carriers. Compatible with following years.

The re-emergence of foreign trade focused on the exchange of raw material exports for manufactured imports. Both required the restoration of effective transport links with other countries, but the former had the greater impact. Major trading partners included countries in eastern and western Europe, and Japan. There was, therefore, much investment in merchant shipping and seaports, the latter especially on the Black Sea (Ilichevsk and Yuzhnyy), on the Baltic (Klaipeda and Liepaya), and in the Far East (Nakhodka and Vostochnyy: Fig. 5.1). Land transport into eastern Europe was also transformed, with high-capacity oil and gas pipelines (Fig. 5.3), high-voltage transmission lines, and enlarged transhipment and bogie-adjusting facilities at the frontier railway stations, where the Soviet 5 ft 0 inch gauge changed to the standard 4 ft 8.5 inch used in all of Eastern Europe except Finland.

For some years the tonnages transhipped at the main border crossing points exceeded those at major Baltic ports. This ceased to be true when railways on the 5 ft gauge were extended into Poland, Czechoslovakia, and Romania (Fig. 5.4). Bulgaria and East Germany, lacking a common frontier with the Soviet Union, were linked to it by long-distance train ferries in 1978 and 1986 respectively (Fig. 5.1).

The continuing desire to exploit domestic raw materials, together with rapidly expanding domestic demand and the addition of foreign markets, placed great strains on transport. In particular the Soviet Union exploited northern resources on a very large scale, and it established a large population in remote and difficult regions in order to do so. Thus where Canada and Alaska, with their more restricted notions of economic accessibility, could make do with light seasonal or air transport, complementing pipelines for oil and gas, the Soviet Union had to provide high-capacity and often all-year facilities. This particularly applied to the oil and natural gas of north-western Siberia and the platinum-group metals of Noril'sk. Pipelines to move oil westwards were one of the main national transport investments in the late 1960s and early 1970s. Thereafter the emphasis shifted to gas pipelines further north (Fig. 5.3). Annual movements westwards by the late 1980s comprised over 250 million tonnes of oil and 350 billion cubic metres of gas. Though river transport carried most of the supplies moving north to the oil and gas fields, over 20 million tonnes a year, it had been supplemented by all-year transport. A railway from Tyumen' reached the Taz peninsula in 1986, and shipments of pipe by sea to the Ob' and Taz gulfs, started in 1979, had grown to over 400 000 tonnes a year.

All-year sea transport had also become vital for Noril'sk, which was shipping over 3 million tonnes of concentrates annually to smelters on the Kola peninsula, in order to avoid new investment and use their surplus capacity. But Noril'sk too received most of its supplies by river from the south. Further east, in the Lena basin and the extreme north-east, exports of diamonds, gold and tin did not require high-capacity facilities. Supplies, however, still had to be brought in. Since there were no railways, and the maritime shipping season was even shorter than that on the rivers, the River Lena became the regional lifeline. In 1951 it was linked to the railway system through Osetrovo, a port which suffers from summer low water and was scarcely able to cope with the ever-growing traffic despite several rounds of investment. Supplementary all-year access to the north-east was provided by roads northwards from the Trans-Siberian railway, and both northwards and westwards from Magadan, an all-year port on the Sea of Okhotsk. Winter driving conditions were among the worst in the world.

Northern resources were also exploited west of the Urals, though under less extreme conditions. They included apatites and iron ore in the Kola peninsula, coal, oil, and natural gas near the Urals, and timber everywhere. Access was primarily by rail, with sea and waterway transport playing a supporting role.

Providing transport for northern resource exploitation was very expensive, but it gave the Soviet Union more experience than any other country in building railways, roads, and pipelines through muskeg and permafrost, and in operating Arctic river and maritime transport.

Pressures occasioned by burgeoning natural resource traffic were not confined to the north. Though partly

supplanted by oil and gas, coal remained in heavy demand and formed the main component of east–west railway traffic west of the Kuzbass. By the 1980s, with the additional burden of forest products from East Siberia, grain from West Siberia, and grain and minerals from northern Kazakhstan, the Trans-Siberian and the parallel railway lines to the south of it had reached a state of chronic overload.

Grain traffic grew rapidly with the implementation of the Virgin Lands agricultural colonization scheme in West Siberia and northern Kazakhstan in the 1950s and early 1960s. Following decades of agricultural neglect in favour of mining and heavy industry, it posed an unaccustomed challenge to transport: how to service a large area, rather than a single point, generating heavy traffic. The solution first chosen was narrow-gauge railways, but later these were converted either to broad gauge or to highways. The east–west broad-gauge segments, when joined together, became the Central Siberian railway (Fig. 5.4).

In the early 1970s it was anticipated that some Siberian resources would move not westwards to Europe, but eastwards to the Pacific coast for export, primarily to Japan. This expectation, together with fears for the vulnerability of the Trans-Siberian railway in an era of tension between the Soviet Union and China, stimulated the revival of a pre-Second World War scheme to build a railway from Siberia to the Pacific, north of the Trans-Siberian. Construction of the Baykal–Amur Mainline (BAM) has continued ever since: one 15-kilometre tunnel remained incomplete in 1993, together with many ancillary facilities. The line was originally supposed to be finished in 1983, but enthusiasm waned when world commodity prices slumped and it became evident that neither the Japanese nor any other Pacific market could absorb all that Siberia had to offer.

One reason for concern about the vulnerability of the Trans-Siberian was its role in supplying the Soviet Pacific fleet. This brings us to the final national policy mentioned at the beginning of this section, namely the expansion of Soviet military activity. That it helped stimulate the building of the BAM is suggested by Western reports of large stockpiles of military supplies along the line. The logistical needs of the Soviet navy also undoubtedly stimulated investment in civilian seaports and merchant ships. This may have been especially true in the Arctic. Much of the heavy investment in ports, icebreakers and icebreaking freighters would be hard to justify on purely economic grounds.

Transport policies and technical innovations. Minimizing investment in transport continued to be Soviet policy

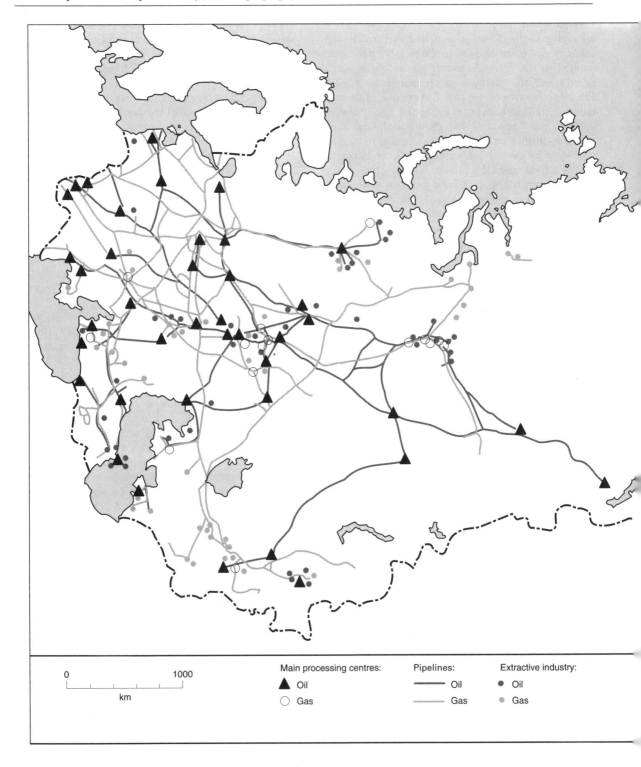

Main processing centres:
▲ Oil
○ Gas

Pipelines:
— Oil
— Gas

Extractive industry:
● Oil
● Gas

0   1000
km

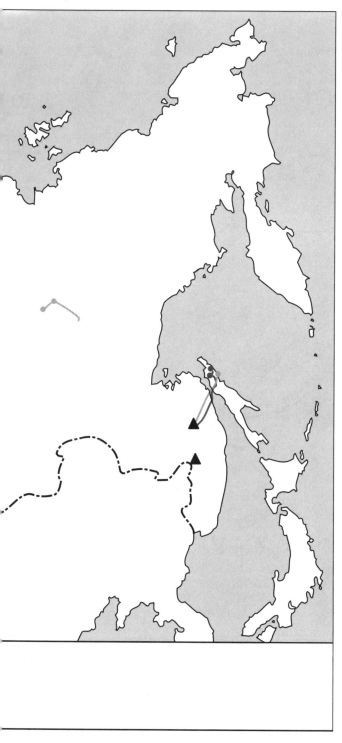

Fig. 5.3
Oil and gas pipeline networks at the end of the Soviet
period.

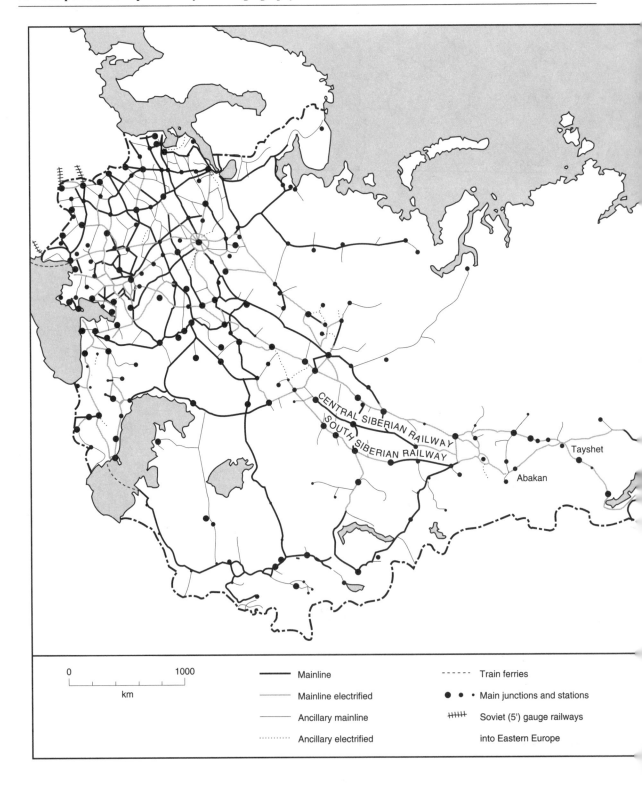

CENTRAL SIBERIAN RAILWAY

SOUTH SIBERIAN RAILWAY

Tayshet

Abakan

0   1000

km

Mainline

Mainline electrified

Ancillary mainline

Ancillary electrified

Train ferries

● ● · Main junctions and stations

Soviet (5') gauge railways

into Eastern Europe

BAYKAL – AMUR MAINLINE

– · – · – ·   Pre-1991 International boundary

Fig. 5.4
The railway network at the end of the Soviet period.

after Stalin's death, but it had less force than before and was no longer expressed in an overwhelming emphasis on railways. Modal diversification was in fact a major trend, with the railways' share of traffic falling by nearly half (Table 5.3). It occurred in part because of uncompromising demands on transport. Overseas trade required sea transport, and there was no real alternative to pipelines for overland oil and gas movements on the scale required. But modal diversification also reflected a recognition that many tasks could be carried out by other modes more efficiently than by railways – an important consideration as economic development shifted from extensive to intensive. Thus passengers were shifted to aircraft or buses, fluids to pipelines, and some low-value bulk freight to water transport. But the process was far from complete by the end of Soviet times, partly because of the sheer scale of investment required. A much greater share of petroleum products continued to move by rail than in North America, for example, because there were higher priorities than building product pipelines. Also there were technical limits to further shifts of traffic from the railways. The rivers were limited by their short navigation season. As for two techniques which might have relieved the railways of much coal traffic, slurry pipelines never got past the experimental stage, and long-distance transmission remained unable to move electricity economically from the Siberian coalfields to European USSR.

Despite its decline in prominence after 1953, the Soviet railway system remained the most heavily used in the world, with over half the world's traffic on little more than 10 per cent of the world's railway network. Furthermore, 50 per cent of that traffic was concentrated on 15 per cent of the Soviet network, and principally along the three major axes.

Costs and prices. We have seen that the Soviet Union developed a transport system with distinctive technical characteristics. No less important a part of the transport legacy to the Soviet successor states were the relationships between costs and transport tariffs. Those relationships were complex. One reason was the vagaries of the Soviet pricing system alluded to in Chapter 4. Few prices in the whole economy were market-determined. Most key prices were set by the state, and they reflected a variety of economic, political and social considerations. It is of course very difficult to calculate the true costs of any operation dependent on arbitrarily-priced inputs, and transport was no exception.

A second reason was transport tariff policy. In any multipurpose transport system, only some costs can be identified with particular movements or types

of traffic. The rest must be assigned according to policy decisions. The main Soviet carrier, the railways, avoided the identification problem to some extent by basing tariffs on network average costs. This had the effect of subsidizing high-cost lines, such as those running north from the Trans-Siberian. There were also overt subsidies within the system. Tariffs for raw material transport, for example, were for long set very low, to encourage resource exploitation far from markets. Movements to the Far East were also subsidized, as was passenger service on metro and other commuter railways.

When transport pricing of this type is adjusted closer to costs, as has happened since the break-up of the Soviet Union, some economic activities find their locations less viable. That is not to say that transport prices dominated the spatial distribution of industry at the end of Soviet times. Many industries were under far more pressure to maximize production than to maximize profits, and for them transport costs were merely an element in their accounts. Others, with sales in the north, were quite happy to be far from their customers, because the government allowed them to calculate profits as a percentage mark-up on delivered costs. Transport operators too could gain most in bonuses by encouraging the growth of traffic – hardly a situation likely to promote either efficient transport or an efficient distribution of economic activity.

Organization. Soviet transport organization was characterized by an allocation of responsibilities strictly along modal lines. The transport ministries both regulated their particular mode and operated it as common carriers. Other agencies operated their own trucks, river vessels, or railway sidings, but only for carrying their own goods. The ministries did not overlap into other forms of transport, with a few exceptions such as the maritime ministry's river fleets in the extreme north-east and on the River Danube. There was never anything comparable to the Canadian Pacific organization in its heyday, operating ocean and coastal shipping, a transcontinental railway, an airline, and a trucking fleet. This situation worked against the Soviet government's stated aim to create a 'unified transport system'. Since investment capital was always scarce, and the transport ministries' plan targets were always expressed principally in tonne-kilometres, the ministries usually functioned as devoted lobbyists for their own modes, and they did not easily cooperate with one another. Poor co-ordination and delays in transhipment were common, and for many years shipments using more than one mode required a separate set of documents for each. Shippers therefore preferred to use only one mode where possible, even at higher cost. Normally they used the railways, the unavoidable mode for part of most journeys for the major freights. Late in Soviet times the government tried hard to promote better intermodal cooperation, but the railways in particular evidently found it difficult to deal with lesser agencies. Especially did they resist the transfer of freight to inland waterways, even from overloaded routes.

The inland waterways in fact can furnish an instructive example of the impact of poor intermodal cooperation on transport geography. Their share of national freight traffic gradually fell, as an increasingly industrial economy became less and less tolerant of slow movements and seasonal interruptions. The waterways came to depend largely on moving sand, gravel and wood, all low-value and low-prestige (though profitable) commodities (Table 5.5). The waterways ministries, whose low prestige was reflected in Union-republic rather than federal ranking, showed determination and ingenuity in identifying more respected (though sometimes less profitable) roles. The Russian Republic ministry made great efforts to show how effectively it could service northern regions, where roads and railways were still expensive rarities. It also developed new types of craft, such as river–sea vessels and hydrofoils. The former could be used both on inland waterways and coastal seas, so that they reduced the need to cooperate with the railway and maritime ministries. They broadened the range of direct, no-transhipment routes which could be offered to customers, and above all they offered the chance to participate directly in foreign trade and earn hard currency. These advantages were reflected in the use of river–sea vessels in roles for which they would probably have been considered uncompetitive outside the Soviet Union.

The rapid development of hydrofoils contrasted with the laggardly evolution of air-cushion vehicles. This also reflected departmentalism to some extent. Hydrofoils clearly fall within the purview of the waterways ministries, but air-cushion vehicles are neither ships, nor land vehicles, nor aircraft. In consequence they were avoided by all ministries for several years, despite their evident potential in north-western Siberia, a priority development area. Only the waterways ministry showed interest, and then only in 'skeg-type' (non-amphibious) designs.

The poor development of motor transport (see Fig. 5.5) reflected, in addition to natural conditions, not so much the organization of transport as that of society in general. In particular it reflected the lack of consumer lobbies and the supremacy of 'planners' preferences'. Motor transport offers convenience and versatility, but its building and maintenance costs are usually very high. In Western countries roads are often subsidized from general government revenue, and the decision to build is a political as much as an economic one. Car and truck owners constitute the necessary powerful lobby. But in the Soviet Union there was nothing comparable, and the railways ministry remained the most powerful transport lobby. Furthermore, Soviet planners had no desire to stimulate small-scale enterprise, often an important attraction of road transport in the West, and they did not seem to rate highly the time savings which are a major benefit of private car ownership in most countries.

Traffic. The combined impact of all the factors affecting transport is expressed in traffic flows and their distribution among the various modes. The volume of Soviet freight traffic grew nearly eight-fold in the 40 years up to 1953 and more than eight-fold in a similar period thereafter (Table 5.3). The dominant role of raw materials in this growth is reflected in the rise of pipelines, which were insignificant in 1953 but carried over one-third of all freight by volume in 1988 (Table 5.3); in the predominance of raw materials and the products of initial processing on the railways, coupled with a steady rise in the average length of haul (Tables 5.1 and 5.2); and in the sharp rise in prominence of sea transport, with a similar commodity breakdown and rise in length of haul as it moved into foreign trade (Tables 5.1, 5.3 and 5.6).

Table 5.5 Waterway freight dispatched, common carrier, by commodity, USSR, 1985 and 1990 (million tonnes, percentages)

	1985 Weight	1985 Per cent	1990 Weight	1990 Per cent
Oil, etc, in tankers	40.0	6.3	34.0	5.1
Wood in rafts	50.0	7.9	32.3	4.8
Wood as cargo	18.7	3.0	18.4	2.8
Coal, coke	20.3	3.2	17.3	2.6
Ore	10.5	1.7	8.1	1.2
Iron and steel	4.1	0.6	3.4	0.5
Grains	6.5	1.0	6.7	1.0
Other	482.5[1]	76.3	548.8	82.0
Total	632.6	100.0	669.0	100.0

Sources: TsSU SSSR, *Narodnoye khozyaystvo SSSR v 1985g.. Statisticheskiy yezhegodnik* (Moscow: Statistika, 1986), p. 332; Informatsionno-izdatel'skiy tsentr Goskomstata SSSR, *Transport i svyaz'. Statisticheskiy sbornik* (Moscow, 1991), pp. 34, 37–9.

Note: 1. Includes mineral building materials (sand, gravel, etc.) 440.0 (69.6%).

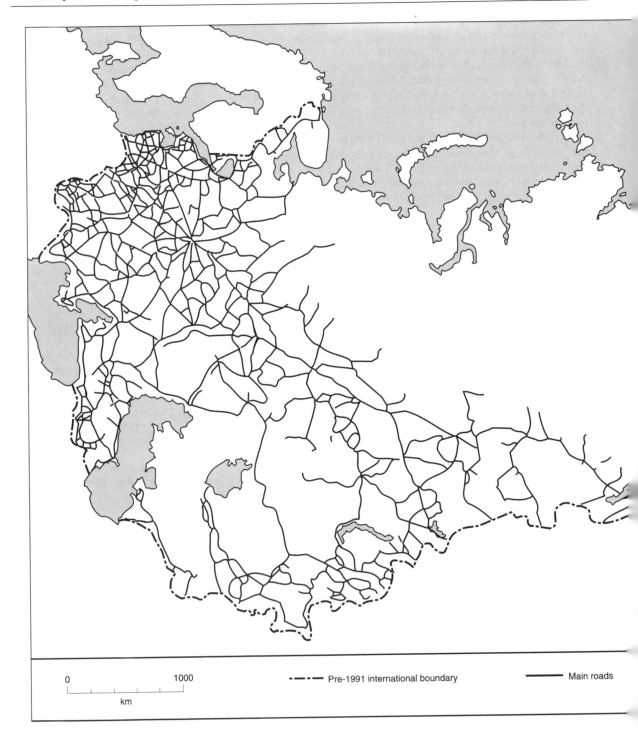

0   1000   ----- Pre-1991 international boundary   ———— Main roads

km

Network of minor roads
(N.E. region only)

Fig. 5.5

The road network at the end of the Soviet period.

Table 5.6 Maritime freight dispatched, common carrier, by commodity, USSR, 1985 and 1990 (million tonnes, percentages)

| | 1985 | | 1990 | |
	Weight	Per cent	Weight	Per cent
Oil in tankers	95.1	39.7	85.4	37.3
Wood	10.1	4.2	9.3	4.0
Coal, coke	10.7	4.5	13.3	5.8
Ore	17.8	7.4	14.6	6.4
Iron and steel	13.5	5.6	8.7	3.8
Grain and milled products	18.3	7.6	18.3	8.0
Fertilizers	5.0	2.1	6.8	3.0
Other	69.2[1]	28.9	72.6	31.7
Total	239.7	100.0	229.0	100.0

Sources: TsSU SSSR, *Narodnoye khozyaystvo SSSR v 1985g.. Statisticheskiy yezhegodnik* (Moscow: Statistika, 1986), p. 330; Informatsionno-izdatel'skiy tsentr Goskomstata SSSR, *Transport i svyaz'. Statisticheskiy sbornik* (Moscow, 1991), pp. 34, 37–9.

Note: 1. Includes mineral building materials (sand, gravel, etc.) 18.9, chemicals other than fertilizers 5.0, machinery and equipment 4.7.

Table 5.7 Waterway traffic, by river basin, 1975 (million tonnes, percentages)

	Weight	Per cent
Volga-Kama	201.5	42.4
Ob'-Irtysh	42.2	8.9
North-west European USSR	41.2	8.6
North European USSR	37.6	7.9
Don – Sea of Azov	20.8	4.4
Yenisey	17.1	3.6
Amur	16.9	3.6
East Siberia (Angara-Baykal)	15.3	3.2
North-east, incl. Lena	9.0	1.9
Non-RSFSR	73.9	15.5
Total	475.5	100.0

Source: Tonyayev, V.I., *Geografiya vnutrennikh vodnykh putey SSSR* (Moscow: Transport, 1977), pp. 47–9.

By the mid-1980s nearly half the total flow in the oil pipeline system, or 250 million to 280 million tonnes per annum, comprised crude oil moving from West Siberia to European USSR and eastern Europe. The pattern of gas movements was similar, while the railways were carrying westwards from Siberia some 50 million

tonnes of coal, mostly from the Kuzbass, and large quantities of wood and grain. There were big freight movements in other directions too, such as chemical fertilizers to Central Asia from the Kola peninsula, coal to the Pacific coast from south Yakutia (now Sakha), and a general distribution of iron and steel from the Dnepr, Donbass, and south Urals industrial regions, but the westerly flow of raw materials was strikingly prominent. As for sea transport, foreign trade accounted for only 12 per cent of traffic in 1940, but 90 per cent in the late 1980s, and the biggest component by far was oil, exported mainly through the Black Sea port of Novorossiysk. It was followed by mineral building materials, carried particularly in the Black Sea coastal trade, and grain, imported from North America, Australia and Argentina (Table 5.6).

The physical and technical limitations of the inland waterways are reflected in their small and declining share of traffic and the predominance of sand and gravel in their freight (Tables 5.3, 5.4 and 5.5). The locational advantages of the Volga and its tributaries, which join complementary regions, i.e. regions lacking and those having in surplus wood, coal and oil, together with the creation of a European deep-water system based on the Volga and joining the White, Baltic, Black and Caspian Seas (Fig. 5.1), resulted in that system carrying over 60 per cent of all waterway traffic (Table 5.7). The only heavily-used river basin not connected to that system was the Ob'-Irtysh, serving the West Siberian oil and gas regions.

A regionalization of transport at the end of Soviet times. From the preceding discussion it is clear that the transport system had reached different stages of development in different parts of the Soviet Union. A basic division can be made between regions within and outside areas of continuous settlement, the latter comprising the north, the deserts of Kazakhstan and Central Asia, and parts of the mountain rim (see also Ch.1). These 'pioneering' regions had high-capacity transport facilities, such as the pipelines serving north-western Siberia and the railways across the Kazakh desert, but they lacked a developed transport network. Parts of the mountain rim had a relatively good road network, designed for frontier control, but for the most part the pioneering regions were served by the branches of a spine-and-branch system, reaching out from the trans-Siberian spine to tap specific natural resources and facilitating movements in only one or two directions, often for only part of the year and certain types of traffic. The Northern Sea route can be regarded as a second east–west spine. During the 1930s it gave primary access to much of the north, with the rivers flowing to the Arctic as its branches. After the

Second World War its importance for primary access declined relative to the railways, but its basic function remained.

Within the continuously-settled area there was normally a transport network, but the quality declined from west to east. Much of the area west of the Urals was well served by a variety of transport modes, facilitating reasonably direct movements in any direction. However, the network was less dense than in eastern Europe and much less so than in western Europe, especially in the case of roads. East of the Urals there was a sharp deterioration to a still more open network, more dependent on railways. East of the Kuznetsk basin (Kuzbass) the term 'network' was barely accurate. Rather there were two spine-and-branch railway systems, touching at Tayshet and occasionally linked elsewhere. Rivers and even roads functioned only as branches.

Post-Soviet transport

The Soviet Union ceased to exist at the end of 1991. Changes in and affecting transport began earlier in response to *Perestroyka* and accelerated after the break-up. Even so, at the time of writing, it is still too soon to speak of new transport systems dominated by post-Soviet influences. The organization of transport and the rules of the game for operators have changed, but the old infrastructure and much of the old pattern of demand remain. The following sections will therefore discuss changes in terms of how they have modified the Soviet transport system, focusing mainly on Russia.

Organization. Dissolution of the Union brought different parts of the Soviet transport system under different national authorities. An immediate effect, following the practice of new countries elsewhere, was the creation of national-flag airlines. These were normally former regional divisions of Aeroflot. In the case of railways, the difficulties of breaking up a highly integrated system were recognized, and attempts were made to continue some form of central organization or at least co-ordination. Maritime transport, on the other hand, was disrupted by a decision to divide up the assets of the Baltic and Black Sea merchant fleets. Fixed assets, such as ports, were deemed to belong to the countries where they were located. Ships were allocated to their ports of registration. But several of the major ports, especially on the Baltic, had handled mainly freight in transit to or from Russia. Also, ports on both seas had specialized in particular cargoes for the whole USSR – Novorossiysk in petroleum, for example. The new order left the Baltic states with

surplus ports and ships, and Russia with too few to handle its foreign trade. Also the Baltic states acquired Arctic-class ships, which were irrelevant to their needs and expensive to operate, while Russia acquired most of the oil tankers which had formerly served all the Black Sea republics. None of this would have caused problems in an atmosphere of mutual trust, but that did not exist.

Other organizational changes began before dissolution. *Perestroyka* brought attempts to devolve operational and financial responsibility from the central ministries to regional transport organizations and even individual enterprises, ranging from railway workshops to ports and shipyards. The transport ministries were supposed to confine themselves to regulation and long-term planning, though they soon found ways to re-assert operational control. In Russia after dissolution, the policy of devolution was continued but modified. Two ministries were created. The Ministry of Communication Routes is responsible for the railways, which remain under full government control. The Ministry of Transport covers all other modes and is mainly regulatory, though with some operational responsibilities which are supposed to disappear after a transitional period. Below the ministry level many organizations have been privatized, though their form and functions in some cases are much as they were under *Perestroyka*, ownership having simply passed to their managers and employees in joint-stock companies. At the end of 1992, for example, all 10 maritime shipping companies, 21 out of 41 ports, and 11 out of 13 ship-repair yards were turned into joint-stock companies. Other organizations have been broken down regionally or sectorally for privatization: Aeroflot has given birth to no less than 70 new airlines. Alongside these privatized organizations there are new private companies. Some, mainly in motor transport, have been set up by individual entrepreneurs. Others are subsidiaries of the privatized organizations or joint ventures with foreign companies. Others again are the creation of regional authorities wishing to assert their independence of the central government.

The level of competition in transport, which used to be more or less restricted to intermodal competition, was in some cases heightened by *Perestroyka* and privatization even before the dissolution of the USSR. One example was on the River Danube, where the Soviet river fleet was run by the Ministry of the Merchant Marine. The new conditions brought competition from the Dnepr shipping company, which successfully sought the business of small shippers. In the past almost all business came from large government concerns, so the small shippers were virtually ignored and the fleet built up no experience in attracting them. In the Far East, however, where the maritime shipping companies found themselves competing with each other for foreign business, they soon tried to form a cartel instead.

Financing. The devolution of financial responsibility was perhaps the most traumatic experience of *Perestroyka* for many transport organizations. Most of them had never had to match revenues with expenditures: the ministries had collected their revenues and paid their costs. Many found themselves unable to meet their costs in the absence of tariff increases. Furthermore, established measures of performance like the maximization of tonne-kilometres, on which pay bonuses had been based, now had to be replaced by measures relevant to the search for profits. In Russia the trauma has become worse since dissolution. Traffic, and therefore revenues, have declined. Prices for many goods needed by the transport companies have been freed and have risen rapidly, but some transport tariffs remain under regulation and have risen much more slowly. In 1992 railway inter-regional tariffs were allowed to rise eight-fold, but the railways' expenses rose between 25- and 70-fold. In addition, the government has stopped paying subsidies even where, as in the north, it is recognized that the economy can hardly function without them. The slow and erratic emergence of a commercial banking system has meant that even the soundest investments cannot easily be financed.

Traffic. Traffic has fallen on all modes except gas pipelines. The decline on some railway routes has been approaching 20 per cent per annum; motor traffic in Russia fell 25 per cent, waterway traffic 41 per cent, rail traffic 15–20 per cent, and sea port turnover nearly 12 per cent in 1992; and the decline continued in 1993.

Two influential factors in the decline have been the collapse of the command economy and the collapse of the rouble (see Ch.4). The former means that suppliers are no longer forcibly linked to specific customers. The latter means that anyone able to do so is likely to sell abroad for hard currency. Good opportunities have arisen for suppliers of commodities, especially metals. Domestic manufacturing industries have therefore been unable to obtain inputs. Another problem for many industries is that their suppliers or customers, or both, are now in foreign countries. Differing monetary policies in the various countries make it difficult to arrange payments for goods and services. Rampant inflation in most of them makes it difficult even within national boundaries. For all these reasons industrial production and traffic have declined. Agricultural traffic has also fallen, partly because

Table 5.8 Freight dispatched, common-carrier railway transport (million tonnes)

	Jan.–June 1991	Jan.–June 1992	1992 as % 1991
Azerbaijan	16.9	12.0	71
Armenia	6.5	1.4	21
Belarus'	56.2	52.3	93
Kazakhstan	171.9	145.1	84
Kyrgyzstan	2.5	2.0	79
Moldova	8.6	5.3	62
Russia	989.3	880.5	89
Tajikistan	3.8	2.9	77
Turkmenistan	15.6	12.6	81
Uzbekistan	42.4	33.0	78
Ukraine	455.6	385.3	85
Total	1769.3	1532.4	87

national and regional authorities, fearing shortages, have intermittently forbidden cross-border movements. For example, in mid-1991 all the Siberian krays and oblasts, and the Sakha Republic, were restricting exports. Strikes, the breakdown of CMEA links, and armed conflict, especially in the Caucasus, have all depressed traffic further (Table 5.8).

Decline has not occurred evenly. Indeed, demand grew on many routes to hard-currency trading partners until 1992, causing congestion in ports and along railways. In the Russian Far East, both Vladivostok and Vostochnyy were clogged with export traffic. Foreign-trade traffic has since declined, though not all routes have been affected equally. Firstly, Russia has been diverting traffic from Baltic state ports like Tallinn, Riga and Klaipeda to those under its own control. Its plan for 1993 was to divert 5.5 million tonnes to St Petersburg, Kaliningrad and Vyborg, and 2 million tonnes to Arkhangel'sk and Murmansk. Secondly, foreign trade in some regions has been stimulated by their inability to obtain supplies domestically. Kamchatka, for example, has had to arrange its own fuel supplies on occasion. Thirdly, new landlocked states have arranged for specific ports to handle their overseas traffic. In the Russian Far East Kazakhstan has chosen Sovetskaya Gavan' and the Central Asian republics, Nakhodka.

Traffic pressure has not been restricted to railways and seaports. In the Far East much of the burgeoning trade with China has been moving either by motor transport or along the River Amur and its tributaries, and the Amur fleet has been handling about 20 per cent of trade with Japan. New international air routes have also been proliferating.

The relative growth of traffic on foreign-trade routes does not simply reflect the traffic offered. Transport companies have undoubtedly favoured foreign trade, because it is much more profitable. Scarce supplies of aviation fuel have been made available for new foreign routes at the same time as domestic flights are being cancelled for lack of fuel.

Passenger traffic, measured in passenger-kilometres, was falling by about 1 per cent per month in 1992 and the first half of 1993. The share of rail transport was rising rapidly: the other modes were suffering more from fuel and equipment shortages and raising their prices more quickly. Domestic air passengers especially were switching to the railways. In 1993 railway passenger traffic in the Russian Federation rose by 7 per cent.

Investment. As the preceding section implies, the money available for transport maintenance and investment has declined overall, but companies serving foreign trade are much better off than those confined to domestic traffic. Especially well off are the maritime shipping companies which carry freight between foreign ports. In 1992 the Russian companies carried 22.7 million tonnes of Russian imports and exports, but 48 million tonnes for foreign customers.

The overall decline in traffic has been fortunate in some ways, because most transport organizations could not have coped with an increase. Standards of maintenance have declined and accident rates have increased on most modes. Motor transport accidents rose between 20 and 30 per cent from 1991 to 1992, despite the 30 per cent drop in traffic, and Aeroflot, or its successors collectively, achieved the unenviable status of having the worst accident record among world airlines. But those maritime shipping companies, airlines and river shipping companies with international routes have been able to earn hard currency, and the Russian government has allowed them to keep some of it. They have been buying ships and aircraft and upgrading ports and airports. At the same time the Kamchatka Maritime Shipping Company, which has little foreign traffic, anticipates shutting down within a year or two for lack of money to replace its aged ships, and local airports in north-eastern Siberia may be closed unless regional authorities or customers take them over.

Company initiatives are being supplemented by some government initiatives, despite a lack of funds. The Russian government has already begun construction of a new port on Luzhskaya bay, west of St Petersburg, one of fourteen it intends to build in the next fifteen years. If these plans go ahead, the Baltic states will be left with a great deal of surplus port capacity, and the recently built port in Estonia, east of Tallinn, will be completely redundant. In the Far East current efforts are focused on building more bridges over the River Amur, to facilitate traffic with China.

Equipment poses serious problems for maritime shipping companies. In the first place most of the vessels acquired in the 1960s and 1970s are due for replacement. Many are too old to be acceptable in foreign ports, which hampers efforts to earn the money to replace them. Secondly, companies operating in the north need specialized equipment. Unfortunately, much of the equipment which was attractive under the old regime is now considered a burden. Icebreakers and icebreaking freighters of the late Soviet era were designed to be versatile and to keep functioning under extreme conditions. That they were very expensive to build and operate was a minor consideration. The costs of operating and replacing them under market conditions are daunting. Furthermore, the Far Eastern companies, unlike the Murmansk Company, cannot use them in the north all the year. The rest of the time they are used on normal trans-Pacific routes, for which they are quite uncompetitive.

The future

Some fundamental influences on future transport development cannot be forecast with any confidence. Much will depend on the outcome of political power struggles among and within national governments, and between national and regional interests. Some Russian politicians would dearly like to re-assert control over the former Union republics, or at least to reunite their economies in a new rouble zone. Either event could reverse the forces now altering traffic flows and transport investment patterns. But strained relations among the former Soviet states and more regional autonomy within Russia will strengthen those forces. In particular, attention will be focused on what in Soviet times were regarded as subsidiary local transport problems, such as the improvement of regional road networks, and on the provision of independent regional outlets to world markets.

Some future influences on transport can be forecast on the assumption that present trends will continue. Unless an impecunious Russian government can restore subsidies, economic activity in the north seems likely to decline. Gold production has already fallen by half in parts of the north-east. This in turn has brought an exodus of population: 18,000 people, or 5 per cent of the population, left Magadan oblast, in 1991, while 55,000 or 30 per cent, left the Chukchi Autonomous Region in 1993. Even where present levels of resource extraction are maintained, the population involved will be much smaller than in Soviet times. Requirements for general-freight and passenger transport will therefore decline. Competition for freight between river and maritime transport will intensify, and in direct competition the former usually has lower costs. But the smaller the traffic volumes, the more vulnerable both will be to competition from air transport. Two factors, however, could reverse the decline in northern activity. The first would be a sharp rise in world commodity prices, the second an overwhelming desire by northern regional governments to earn hard currency. Bereft of subsidies, they may find it hard to resist selling off their natural resources.

Other transport developments can be forecast on the grounds that they can scarcely be avoided in anything other than an isolated command economy. One is an enhanced role for motor transport and much heavier investment in roads, especially rural roads. Intercity highways may be less necessary than in western Europe and North America, but the missing stimulus to their creation in Russia, a sufficiently powerful road-users' lobby, will probably result from the privatization of road transport. The fate of road transport in poverty-stricken Central Asia and in those republics with less commitment to private enterprise is not so easy to forecast.

Another necessary development is the emergence of true intermodal transport companies, as opposed to modal companies which happen to use other modes for internal purposes. Soviet transport was plagued by poor intermodal co-ordination and a predilection for thinking in terms of how to employ a given body of transport equipment, rather than of the best way to move a given cargo. A major stimulus to intermodal thinking in Russia may be cooperation with foreign companies in trying to make Russian facilities more attractive for transit freight. At present this is focused on the revival of the Trans-Siberian Landbridge, but other possibilities include the recently completed transcontinental railway through Russia, Kazakhstan and western China, and the Northern Sea route.

The reference to transit routes leads to the final type of forecasting, based on assessing the prospects of visionary or fanciful projects. Among many old ideas to re-emerge after the break-up of the USSR is an international scheme for a railway tunnel under the Bering Strait. It would seem insupportable on any analysis of potential traffic and ability to compete with trans-Pacific sea transport. Even less likely to be realized in the foreseeable future are the various Soviet-era schemes for east–west railways in northern Siberia, submarine freight transport under the Arctic ice, and a combination of airship and helicopter technology known as the vertostat. For a considerable time, financial exigency is likely to restrict transport developments in the former Soviet Union to what is severely practical and relatively cheap.

6

Demographic and social problems

R.A. French

Contrasting patterns of demography

The successor republics of the USSR are faced with a formidable complex of problems in establishing their future political and economic life. Many of these problems are caused by, and most are exacerbated by, the demographic legacy of the Soviet period. Throughout the Soviet Union's existence, every aspect of the country's life and economy was deeply affected by a series of population catastrophes – the First World War and the civil war after the revolution, the Second World War and the appalling losses of the Stalinist regime and its policies. It is estimated that in the period from 1914 to 1923, the country lost some 18.5 million through war casualties, civilian deaths from disease and famine, and emigration, to which must be added a birth deficit of some 10 million. The number of deaths from the Stalin regime and from the German invasion cannot be disentangled, as a result of the publication of false, exaggerated figures for the 1939 census, but the total losses between 1928 and 1953 from both these disastrous causes may be estimated as between 22 and 28 million. These disasters had not only heavy immediate impacts on population numbers, but also on-going consequences which have affected the demographic structure down to the present and which remain as underlying problems for the post-Soviet republics.

In the first place, the losses were borne principally by the male population, leaving a serious sex imbalance. At the time of the 1959 census, fourteen years after the end of the Second World War and six years after the death of Stalin, there were 20.7 million more women than men; by the most recent census of 1989, the imbalance had only reduced to 16 million; males formed 47.6 per cent of the total population. This disproportion has always meant a crucial need for women to play a full part in the labour force, a need reinforced by the Marxist belief that women

had a right to work and thus to escape the constraints of domestic exploitation. The necessity for women to work has been further strengthened by the economic pressure for families to have two bread-winners, a situation which the economic chaos of the post-Soviet period has powerfully reinforced.

In the 1990s, the excess of females is seen only in the older age groups. The passage of 40 years since the Stalin terror and the normal demographic situation of a preponderance of male births has meant that by the time of the 1989 census, there were nearly 2 million more males than females under the age of 30, although only 230 000 of this excess were in the 20–29 age groups. Among those in their thirties and forties there were 1.3 million more women than men, but over the age of 49 the disproportion increased very sharply, with a surplus of almost 16 million women. The imbalance is slowly working its way out of the age pyramid, but the labour-intensive economy bequeathed by the USSR and the abiding, indeed increased, need for pension supplementation mean that older women, including those over the official retirement age of 55, very commonly continue to work.

This imbalance between the sexes is recorded in every republic in the older age groups, but is distinctly lower in Central Asia. The proportion of males in the over-40 population of Tajikistan is 47.5 per cent; the equivalent figure for Ukraine is 40.3. This, of course, reflects the greater losses in Ukraine and in other parts of European USSR, which were occupied by the Germans and fought over in both World Wars, and which underwent the severest famine effects of the collectivization programme in the 1930s. At the same time, the Central Asian republics have a far smaller proportion of older people, in consequence of the higher birth rate, and also a smaller proportion of women who participate in the labour force. In these republics, as a result, the imbalance of the sexes is of relatively minor significance.

As in the case of the sex balance, the demographic trends since the Second World War, while everywhere showing the influence of the war losses, display two markedly differing patterns, which one might loosely term 'European' and 'Asian'. However, in this context, the Siberian and Far Eastern part of the Russian republic can only be considered as part of Europe. Its inhabitants are overwhelmingly Russian; non-Russians form less than 4 per cent of the population. A third group of republics, in demographic as in other social matters, might be termed 'Intermediate'. The 'European' group includes the three Slav republics, Russia, Ukraine and Belarus', and the three Baltic republics, Estonia, Latvia and Lithuania. In all of them, the rate of crude natural increase, excess of births over deaths in any given year, is below five per 1000 (Table 6.1). The 'Asian' republics of Azerbaijan, Uzbekistan, Kyrgyzstan, Turkmenistan and Tajikistan, have annual increases of over 20 per 1000 and, in the last instance, over 30. The remaining 'Intermediate' republics have rates of increase between eight and sixteen per 1000.

All former Soviet republics have witnessed a fall in fertility during the postwar period, though the 'Asian' group began their fall from a far higher starting point and have levelled out higher. In the 'European' group, birth rates were fairly high during the 1950s, as the republics, and indeed the whole USSR, experienced a postwar 'baby boom'. This was followed in the 1960s by a steep decline in the birth rate, as the greatly reduced cohort of wartime children entered their twenties, the period of maximum fertility. At the all-Union level, birth rates fell from around 27 per 1000 in 1957 to 17 per 1000 in 1969. There was no corresponding recovery in the 1970s as the larger 'baby boom' generation entered its twenties. By then, the population as a whole was experiencing a fall in fertility. The Slav and Baltic republics have continued with a relatively low level of fertility, especially in the latter case, with birth rates never rising above eighteen per 1000 and only in occasional years in Russia and Belarus' ever surpassing seventeen.

By the end of the 1970s, the Soviet government was increasingly concerned at the fall in numbers entering the workforce, a fall entirely experienced by the 'European' nationalities, that is to say those most mobile, most readily engaged in industry and most needed in the resource-rich, but underdeveloped, regions of Siberia and the Far East. This concern led to a number of pro-natalist measures – maternity grants for the first, second and third children, increased maternity leave and extra days annual holiday for mothers. A slight, though fluctuating, rise in birth rates during the 1980s was probably due to these efforts, but it was short-lived and by the end of that decade, birth rates were once again falling quite steeply, to 13.4 per 1000 in Russia in 1990 and 12.7 in Ukraine. A number of factors combine to maintain the lower fertility in most of the republics. As elsewhere in the world, rising living standards have improved knowledge, availability, quality and use of contraceptive methods and at the same time have increasingly presented material possessions as an alternative to larger family size. But even in the most advanced republics, the better material conditions of life have been only sufficient for limited effect. Rather, it is the deficiencies in quality of life which have proved the most significant influences in falling rates of reproduction.

In the first place, accommodation has been a major brake on increased family size. By the time that Stalin died in 1953, his neglect of adequate housing construction had reduced the average amount of living space per head of urban population to below the pre-revolutionary level, to a mere 4 m². A family occupying only a single room and sharing kitchen and bathroom in a communal flat was the normal situation, rather than the exception. In 1957, Khrushchev launched a colossal housing programme, which continued to the end of the Soviet era. Most of the population of the country have been rehoused, but the elimination

Table 6.1 Birth and death rates, 1989 (per 1000)

Republic	Birth rate	Death rate	Crude natural increase	
USSR	17.6	10.0	7.6	
Russia	14.6	10.7	3.9	
Ukraine	13.3	11.6	1.7	
Belarus'	15.0	10.1	4.9	'European'
Estonia	15.4	11.7	3.7	
Latvia	14.5	12.1	2.4	
Lithuania	15.1	10.3	4.8	
Moldova	18.9	9.2	9.7	
Georgia	16.7	8.6	8.1	'Intermediate'
Armenia	21.6	6.0	15.6	
Kazakhstan	23.0	7.6	15.4	
Azerbaijan	26.4	6.4	20.0	
Kyrgyzstan	30.4	7.2	23.2	
Uzbekistan	33.3	6.3	27.0	'Asian'
Turkmenistan	35.0	7.7	27.3	
Tajikistan	38.7	6.5	32.2	

Source: Narodnoye khozyaystvo SSSR v 1989g., 1990, Moscow: Goskomstat.

of the communal flat is still far from complete and even the new apartments are small. Approximately half those built since 1957 have two rooms, a quarter have only one room and a quarter have three rooms; less than 4 per cent have more than three rooms. Under these circumstances, it is not surprising that lack of adequate accommodation is the second most common reason given in surveys of Soviet women for not wishing to have more children – the first being material difficulties.

Another factor has been the use of abortion as the commonest form of birth control. Legalized on demand in 1926, when the emphasis was on the breakdown of the family as the basic unit of society, abortion was prohibited other than on strict medical grounds in 1936, but was once more made legal on demand in 1955. Now the national norm is for a woman to have six–eight abortions in her lifetime. In 1990 there were almost 6.5 million abortions, 134.2 for every 100 live and still births; this was a drop on the annual rate of over 7 million through the 1980s (*Narodnoye khozyaystvo* 1991:251). This high frequency significantly increases the risk of subsequent still births and thus additionally lowers fertility.

Health in general can be a negative factor for fertility and this has been worsened by growing levels of alcoholism, principally among men, but also to some extent among women. In various surveys, a number of women gave as their reason for not wishing to have more children than they already had, the fact that their husband drank. In a large number of cities, especially those engaged in metallurgical, coal mining and chemical industries, environmental pollution, particularly of the atmosphere, has had serious detrimental effects on health. In over 300 cities, maximum permissible concentrations of toxic pollutants in the atmosphere are exceeded, not un-usually up to several tens of times during the course of the year and by up to 400 times on occasional days. This has played a part in the high levels of chronic illnesses, especially of the respiratory system, birth defects and infant mortality; having slowly fallen to 22.7 per 1000 by 1971, infant mortality nationally rose during the 1970s to a point when data were no longer published. Figures were only published once more when, in the 1980s, they recommenced falling; by 1990 infant mortality had dropped to a level of 21.8 per 1000, a figure that was still very high by the standards of most advanced countries.

A further consideration affecting fertility is the level and quality of service provision of all types, including health and pre-natal services. Most important perhaps of all the factors is that the overwhelming majority of women in the 'European' republics, in all child-bearing age groups other than the youngest (15–19), hold full-time employment. In urban areas, it is rare for women not to have jobs, unless affected by health or full-time study. Therefore, provision of pre-school baby and child-care services is essential. In this respect, the USSR was in general well provided, compared to most other developed societies, but there are still not enough places even in the best served Russian and Baltic republics to allow all working women in urban areas to leave children below school age; in 1990 there were pre-school places for 65 per cent of urban children of appropriate age. Moreover, even when pre-school facilities are available, the quality of care is not always sufficiently good to encourage mothers to leave their children.

The problem of raising children, while holding down full-time work, has often been referred to as the double burden on Soviet women. During the 1980s, there were increasingly frequent calls for greater availability of part-time working and time-sharing, but by the end of the Soviet regime very little had been done to provide more part-time jobs. In fact, rather than a double burden of responsibilities, it is fairer in most cases to ascribe a triple burden to the working-age female in the Soviet Union *and* in its successor states. It is still the norm in most families for women to be left with most of the domestic chores – washing, cleaning, shopping, taking children to and fetching them from pre-school groups or school. In all these tasks, women have usually lacked the same degree of help from efficient consumer durables as in Western countries. Even greater time commitment, compared with Western countries, is required by shopping. Throughout the Soviet period, the shortage of supplies has always meant much trailing from shop to shop in search of goods to buy; then, when the desired goods are found, the system of three queues, at counter, at cashier and at counter again, makes yet further demands on time. By 1994, goods, including Western products, were appearing in shops of the larger cities, but inflation was taking them out of reach of all but the wealthiest; in consequence, ordinary citizens were still condemned to long searches for affordable necessities. It is scarcely surprising that surveys have found that women, both in town and country, have far less leisure time than men.

In contrast to the declining 'European' rates of reproduction are those of the four Central Asian republics and Azerbaijan in the eastern Transcaucasus. These republics shared in the general fall in birth rate during the 1960s, for the same reasons as elsewhere. However, the initial levels were much higher than in the 'European' type of demographic stucture and the fall stopped at levels still much higher. Since then, birth

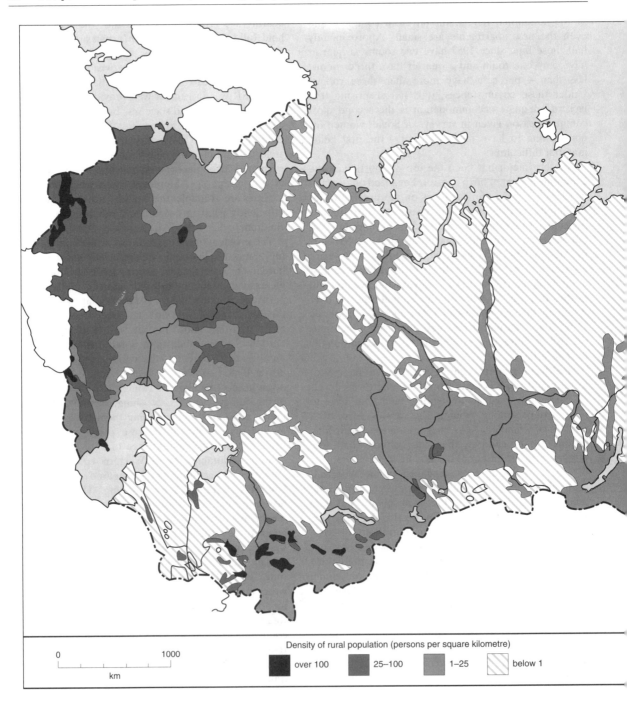

Density of rural population (persons per square kilometre)

over 100   25–100   1–25   below 1

0   1000

km

rates have remained high and by the opening of the 1990s in some republics were showing a tendency to rise, most notably in Tajikistan.

As in the 'European' republics, the 'Asian' birth rates are markedly higher in rural areas than in the towns, but unlike the predominantly urbanized 'European' population, the Central Asians are still more rural than urban dwellers. Islam remains a significant force in Central Asia and traditional cultural attitudes to the role of women in society and to family size are maintained. The proportion of women in employment outside the home is far lower than in

Fig. 6.1
Former USSR: geographical patterns of population density.

the 'European' republics. In countryside and town alike, the extended family unit remains far more usual; this permits those mothers who wish to work full-time to do so, while elderly relatives look after the children. For this reason the indigenous people show a strong preference for the old, traditional type of town house, comprising a group of buildings, with large rooms, attics and cellars, arranged around an interior courtyard, all well adapted to extended family living, rather than the modern, equipped apartment of one to three rooms in a high-rise block.

The higher birth rates of the 'Asian' republics are

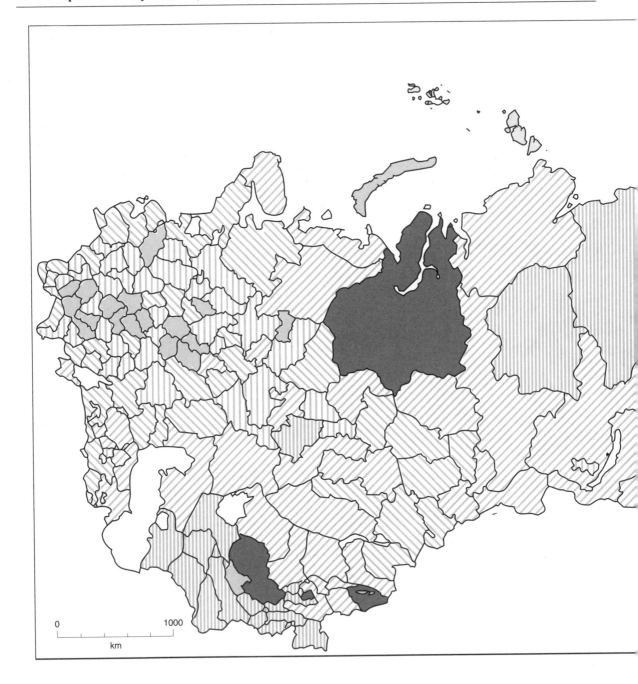

sufficient to outweigh the rates of infant mortality, which are exceptionally high, exceeding 40 per 1000 births in 1990 in Tajikistan and Turkmenistan. Largely these high rates are the consequence of lower levels of health, ante-natal and post-natal care, but heavy pollution of ground water supplies by over-use of chemical fertilizers, pesticides and defoliants on the irrigated cotton fields, has made a major contribution to the mortality.

The 'Intermediate' group of republics includes Moldova, Georgia, Armenia and Kazakhstan. Georgia and Moldova are characterized by a less urbanized population than the 'European' republics, especially in the latter case, and by higher rates of reproduction,

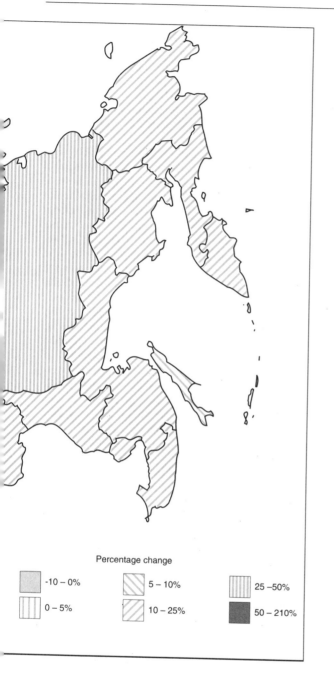

Percentage change

-10 – 0%	5 – 10%	25 –50%
0 – 5%	10 – 25%	50 – 210%

Fig. 6.2
USSR: Population change, 1979–89.

though still well below 'Asian' rates. Armenia combines a level of urbanism above the all-Union average, with a quite high reproduction rate. The last of this group, Kazakhstan, derives its intermediate position from the near balance of Russians and Kazakhs in its population, with each of these groups displaying a demographic structure close to that of their ethnic kin.

Patterns of migration

The differing ethnic groups are also related to differences in the distribution and movement patterns of population. Details of population distribution in the former USSR have already been given in Chapter 1 where the considerable variations in density across the territory were noted (see also Figs 6.1–6.2). Throughout the Soviet period, efforts were made to achieve some degree of redistribution of population. Marxist principle required an even level of development, rather than an exploitive core and exploited periphery. Economic development was seen in terms of opening up the abundant resources of the underpopulated regions, especially Siberia. Geopolitical fears of China's historic claims to the southern Far East and the Amur basin, reiterated in the post-Stalin period of Sino-Soviet disputes, reinforced Soviet aims of populating and developing these borderlands. In the resultant shifts of population, ethnic differences have been sharp. Russians have always been the most mobile nationality, followed by the Ukrainians and, to a lesser extent, the Belorussians, Armenians and the Tatars. In contrast, the peoples of Central Asia have demonstrated a high order of immobility, with few moving outside their own republic and quite negligible numbers moving to other republics outside Central Asia.

Thus for most of the Soviet era, the dominant regional migration consisted of Russians and Ukrainians moving from west to east. In the Stalin terror, the large-scale use of millions of prisoners as forced labour in the GULAG camps, reinforced the eastward shift, as a high proportion of the camps were located in the far north and Far East and those who survived camp sentences to be released were usually required to settle in eastern regions. Free labour in the east was fixed by regulations which required workers to obtain permission to leave from their place of work; unsurprisingly, such permission was not usually forthcoming. Under Khrushchev, the movement eastward was accelerated by his Virgin and Idle Lands campaign (1956–60); some 300,000 young Russians were brought into Kazakhstan and the south of West Siberia to set up new state farms and plough up the eastern dry steppe. But under Khrushchev also, the prison camps were largely emptied and the Stalinist regulations, which tied workers to their place of work, were abrogated. In consequence, the 1960s witnessed, for the first time, a reversal of the eastward flow of population. In many areas of the Urals, Siberia and the Far East, out-migration exceeded in-migration and in a few oblasts, notably Sakhalin Island, the net migration loss exceeded the rate of natural increase and there was an absolute fall in population.

The Soviet government, concerned by the outflow, developed the use of money incentives – regional wage bonuses for the far north, Far East and other areas of difficult living conditions. With inducements by managers, desperate for labour, added to the official increments, workers in these regions could earn five to ten times as much as for the same work in central districts. At the same time, major new economic projects were developing in Siberia, in particular the opening up of the oil and natural gas fields in the West Siberian lowland and the construction of the Baykal–Amur Mainline (BAM) in eastern Siberia and the Far East. Thus in the 1970s and into the 1980s, the pendulum swung once more to a predominant eastward flow.

However, the migration to the east induced by high wages was very largely made up of young, unmarried and unskilled workers. Living very cheaply in hostels, eating in canteens and enjoying free entertainment in enterprise clubs, these young migrants were able to accumulate savings rapidly. At the same time, living conditions were often hard, indeed primitive, while service provision and availability of goods were nearly always poorer than in the European parts of the country. The result was a very rapid turnover of labour, with the majority of migrants leaving again within three years. In particular, it was hard to persuade families to settle permanently in the east; the additional costs of clothing and food (the long, cold winter required a larger intake of calories) offset in part the advantage of higher wages. Lower levels of service provision were particularly manifest in school and pre-school facilities. Siberia and the Far East therefore remained, and still remain, regions of chronic labour shortage and, especially, of a lack of skilled labour.

By the late 1980s and the period of *Perestroyka*, the rate of westward out-migration from Siberia was once more rising, but by then new migration movements were gathering momentum. For most of the Soviet period, the mobile Russians had moved in considerable numbers into all the Soviet republics; unlike the fluctuating situation in the east, the whole period after the Second World War saw a steady movement of population, chiefly Russian, into the western and southern republics of the Baltic, Belarus' and Ukraine. In consequence, the size of the Russian minorities in a number of other republics increased signifiantly. As noted in Chapter 3, the 1989 census showed that Russians constituted significant minorities in Ukraine, Estonia and Latvia (Table 6.2; see also Table 3.1 and Fig. 3.2).

Similarly, absolute numbers of Russians were increasing in two of the southern republics in Asia, Kazakhstan and Kyrgyzstan, although the higher birth rates of the indigenous peoples was causing the proportion of Russians to fall, to 37.8 and 21.5 per cent respectively. However, by the 1980s, growing nationalist attitudes, which became expressible with *Glasnost'*, began to cause Russians to migrate back to the Russian republic. This trend was first evident in the

Table 6.2 Russians as a proportion of total population in non-Russian republics

Republic	1959	1989
Ukraine	16.9	22.0
Belarus'	8.2	13.2
Estonia	20.1	30.3
Latvia	26.6	34.0
Lithuania	8.5	9.4
Moldova	10.2	12.9
Georgia	10.1	6.3
Armenia	3.2	1.6
Azerbaijan	13.6	5.6
Kazakhstan	42.7	37.8
Uzbekistan	13.5	8.3
Kyrgyzstan	30.2	21.5
Turkmenistan	17.3	9.5
Tajikistan	13.3	7.6

Transcaucasus; in Georgia, each successive census from that of 1959 saw an absolute as well as relative decline in the number of Russians. In the intercensal period 1970–79, the absolute decline spread to Azerbaijan; by 1989, it was occurring in Armenia, Uzbekistan, Turkmenistan and Tajikistan.

The outbreak of hostilities between various nationalities, which accompanied the break-up of the USSR, accelerated the process of return to ethnic 'homelands'. The bitter fighting between Armenians and Azerbaijanis over the Nagorno-Karabakh Oblast and the pogroms against Armenians in Sumgait in 1988 have led to the exchange of the vast majority of the Armenian and Azerbaijani minorities in each republic back to their 'homeland' republic, a movement of perhaps 500 000 people. The strife in these two countries has been matched by the civil war situation in the western Caucasus, where Abkhaz and Osetians have been fighting for independence from Georgia. East of the Caspian, armed clan conflicts have taken place in Tajikistan. The hostilities in all these areas have greatly accelerated the outflow of Russians from these republics.

Already the return migrations of Russians, whether forced or voluntary, have placed a series of stresses on all the republics. In Central Asia, the Russians had fulfilled a heavily disproportionate share of industrial and administrative functions; not infrequently they had formed the majority of the population in towns. Now there is an urgent need to induce indigenous peoples to move into the towns and to train them to take over the work previously done by Russians.

At the receiving end, official estimates for December

1992 put the number of Russian refugees in the Russian republic at some 500 000, of whom 150 000 had not found jobs. Other sources have claimed that the true figure may be as high as 2 million. The 'returners' are putting severe additional pressure, not only on the job market, but also on the housing supply. The Herculean efforts made since 1957 to catch up with adequate housing provision, which by the mid-1980s seemed to be within sight of success, have been seriously set back by this influx of returning Russians. The housing situation has been made still worse by the return of large numbers of Soviet troops from East Germany and eastern Europe, and by the breakdown in the financial resources for more new building.

Patterns of social conditions

Throughout the existence of the Soviet Union, data relating to social groups of the population were exceptionally rare. Very seldom were figures of any kind, even demographic, published for ethnic groups; it has been necessary to use Union republic figures as an approximate surrogate. Where social differences are concerned, on the Marxist basis of relationship to the means of production, only two classes were officially recognized – workers and collective farmers. In practice, Soviet sociologists had long been aware of the need for a more sophisticated division of society and had termed such divisions 'layers', but these were never the basis for any publication of statistics. Nevertheless, groupings of society, differentiated by income, privileges, occupation and life styles have existed and do exist and display geographical patterns of distribution.

One of the sharpest contrasts in social geography is, as it has always been, that between town and country. Average incomes of collective farmers have consistently been well below those of workers in factories and offices. In 1990, the average monthly income of workers in industry was 296.2 roubles, in agriculture other than the collective sector 278.9 roubles, and in the collective sector 241.1 roubles (Narodnoye khozyaystvo 1991: 36–8). In terms of almost every statistical yardstick in official Soviet figures, people engaged in agriculture were worse off. The pattern of rural settlement over most of Russia is one of very small hamlets and villages and in the Baltic republics of dispersed farmsteads. This has made adequate provision of all services such as schools and hospitals difficult and expensive. Attempts to ease the situation, by concentration of the rural population into enlarged settlements, designated as growth centres, have met with very limited success (see Ch. 7). Where, as in the steppe regions further south, villages are much larger, they are commonly strung out for many miles along valleys near the water supply, again making access to services difficult.

Of major consumer durables in 1990, only motor-cycles, bicycles and sewing machines displayed higher per capita ownership in rural areas than in towns (*Narodnoye khozyaystvo* 1991: 142). The first two items were the result of the near complete lack of public transport in most rural areas, the last reflected the necessity for women in the countryside to make a higher proportion of their clothes. Compared with 1914, quality of life has improved more rapidly in urban areas and, far from achieving the Marxist aim of abolishing the distinction between town and country, the Soviet period widened the gap. One consequence of a lower standard of living in rural areas is the steady outflow of population, principally the young and best qualified, who can earn so much more in industry (see Ch. 7).

If urban areas enjoy better material conditions in general, there is considerable variation between cities. In broad terms, quality of life deteriorates down the scale of population size. From the 1970s on, planning for urban growth was supposed to reflect a logical hierarchy of size and function, the so-called General Scheme for Settlement. In practice, what developed was a polarization of urban places, with less than 600 medium and large cities standing in a sea of over 5600 small towns and urban districts with less than 50 000 inhabitants (Table 6.3). No less than 3777 urban settlements have under 10 000 population, of which 1301 are under 3000. Despite the efforts to limit growth of the largest cities by the requirement

Table 6.3 Number of urban places by size, 1989

Population	No. of places
Over 1 000 000	23
500 000–1 000 000	33
250 000–500 000	78
100 000–250 000	162
50 000–100 000	278
20 000–50 000	720
10 000–20 000	1 145
5 000–10 000	1 515
3 000–5 000	961
Under 3 000	1 301
Total places	6 216

Source: Narodnoye khozyaystvo SSSR v 1989g., 1990, Moscow; Goskomstat, p. 25.

of a police permit (*propiska*) to reside in them, they have continued to grow, while the smallest places are either stagnating or declining in size. A common cause for decline is the ending of the useful life of a mineral deposit, the working of which was the sole source of employment. Other places have only a single industry, which may be employing predominantly one sex, giving limited opportunity for both partners in a family to find work. Perhaps the greatest reason for the on-going haemorrhage of population from the small town is the very low level of service provision of every type. Indeed, the smallest towns display little difference in this respect from the villages.

In the same way as between towns, disparities in social conditions and their evaluation are found within cities. For much of the Soviet period, social distinction was only very weakly reflected by area of a city, but rather by the date and quality of the housing. The usually well-built apartment blocks of the Stalin period, frequently of granite or other stone, with quite often three, four or five large rooms per apartment, were allocated to the élite, the so-called *nomenklatura*, of senior party members, army and KGB officers, senior management and outstanding figures of the performing arts and academic worlds. They have continued to occupy these buildings. Next in the social order were the cooperative blocks, built by groups which put up the money or took out mortgages; sites were allocated by the local authorities and the completed apartments were owned by the members of the cooperative. The need to put up finance restricted these apartments to better-off people. Ownership causes these buildings to be better maintained and their grounds more carefully tended than the flats belonging to the local authorities.

The vast majority of urban dwellings are in flats built and maintained, either by the local council (*soviet*) or by individual enterprises and organizations. In the desperate housing shortage, new and expanding enterprises had built housing for their workers. In the 1980s government directives required local authorities to take over all responsibility for public housing, but the process of doing so was extremely slow. Frequently, enterprise housing was in sore need of major repair and the local councils were unwilling to accept responsibility before this had been done. Broadly speaking, quality of public housing was related to date of construction. The prefabricated, five-storey blocks of the Khrushchev period, while doing invaluable service at the time to rehouse millions from appallingly overcrowded conditions, are nowadays themselves in need of replacement. The likelihood of this happening in present conditions of economic stress and overstretched construction capacity is slight during the foreseeable future. Later and better constructed prefabricated apartment blocks have been much more highly regarded, but they are beginning to witness the social problems experienced by high-rise dwellings in Western countries, such as loneliness, alienation and vandalism.

Generally low in the social hierarchy is private housing. Most of this is in one- or two-storeyed, wooden houses. Much dates from before the 1917 revolution, or from the period immediately after the Second World War, when private construction was encouraged to assist the task of reconstructing the heavy war losses. Sometimes such houses represent villages, engulfed in the outward spread of urban areas. Not infrequently, provision of mains services in such property is inadequate and it is the home of the poorer layers of urban society. At the very bottom of the ranking of dwellings are the communal flats, where several households, usually in only one room each, share kitchen and bathroom. Frequently conditions in the flats are appalling.

The differences in housing have not necessarily meant corresponding differences in areas of a town. Housing of all classes may well stand side by side on the same street. While the housing shortage remained gravely acute, it was difficult to move and those allocated housing were only too glad to receive it, regardless of location. However, as the stock of accommodation steadily improved in the decades after 1957, so people increasingly sought to move to areas of a city which they perceived to be more desirable. The attractiveness might be in terms of journey to work; locations close to nodes of public transport have always been much sought-after. Certain sectors of a city were seen as environmentally more agreeable, free of industry and pollution, with more parks, nearer to cultural attractions. Gradually, through the process of apartment exchange, definable urban areas have emerged as desirable. In Moscow, such areas include the inner region just west of the Kremlin and two wedges extending right out to the city limits in the north-west and south-west (Sidorov 1992); typically, the south-western sector is almost wholly devoid of industry, is on the upwind side of town and thus little polluted, and has an employment structure dominated by research and development, education and services. It has been common for undercover payments to be made to get into a desirable area, and thus it is the better-off who have slowly begun to concentrate in such areas.

The break-up of the Soviet Union brought about the start of privatizing housing. Occupiers could register a claim to their homes for a nominal sum; although many hesitated for fear of taxation on their property, the number taking out title has gradually grown. The result

is that those who previously occupied the best housing, whether through money, power or special privilege, now own that housing; as a money-based housing market develops, the old élite will be confirmed both in their strong financial position and in their occupation of the most desirable districts.

Just as certain urban areas are seen as 'good' and are thus sought out by the wealthier and more influential members of the community, so other parts are regarded as undesirable. These areas are often the most heavily industrialized sectors, close to sources of air and water pollution and thus with poor health conditions; housing is often old, even pre-revolutionary, and quality of life in most respects is low. Crime rates tend to be higher. Very commonly, new in-migrants tend to congregate in such districts, which are by and large those least favoured by the town's long-standing residents. Newcomers also commonly move into the outer micro-regions, less sought-after because journeys to work and to retail outlets tend to be longer. Also, the pressure to provide housing has always meant that new micro-regions of housing are built and occupied long before the appropriate infrastructure of services and shops is complete; it may well be years before norms of service provision are met in newly constructed parts of a town. In particular, transport services may take some time to extend out into the newer suburbs. The newcomers arriving in these less favoured districts often have trouble adapting and are less stable. A number of studies have demonstrated that there is a clear correlation between the degree of population mobility and crime. Often the outer micro-regions have higher crime rates, reinforcing the perception of these areas as undesirable.

Finally, but perhaps most significantly for the future, the Soviet period left an inheritance of inequality of life style and wealth between the republics. A consistent pattern was established of a north-west to south-east gradient of indices, with the 'European' republics regularly coming at the top of the scale; Estonia or Latvia was commonly in first place (Table 6.4). Equally regularly, the 'Asian' republics came at the bottom, with Tajikistan in most cases taking last place. Such indices include those already referred to, such as provision of doctors and hospital beds per 10,000 population, provision of pre-school facilities, infant mortality, average income of industrial and farm workers, level of urbanization, but also others such as proportion with higher education, quantity of living space per capita, savings and retail turnover per capita.

The former Soviet republics, emerging as independent countries, have all inherited a complex of social problems, but for some the problems are graver than

Table 6.4 Selected social indices, 1989

	A	B	C	D	E	F	G
USSR	44.4	132.9	22.7	240.4	200.8	108	1406
Russia	47.3	138.7	17.8	258.6	221.3	113	1548
Ukraine	43.9	134.7	13.0	217.7	184.3	104	1326
Belarus'	40.6	135.4	11.8	227.8	211.7	108	1559
Estonia	48.3	121.7	14.7	270.1	317.6	117	2164
Latvia	50.0	147.1	11.1	249.9	264.2	115	2055
Lithuania	45.7	125.8	10.7	244.1	257.3	106	1803
Moldova	40.1	126.9	20.4	200.6	196.5	87	1273
Georgia	58.5	110.0	19.6	197.7	169.9	151	1218
Armenia	42.7	90.2	20.4	219.9	204.9	138	1269
Azerbaijan	39.0	99.9	26.2	179.0	182.3	105	814
Kazakhstan	40.9	135.6	25.9	233.6	210.3	99	1168
Uzbekistan	35.8	123.1	37.7	193.8	165.0	92	825
Kyrgyzstan	36.6	119.3	32.2	197.5	198.0	94	963
Turkmenistan	35.5	110.6	54.7	221.3	203.9	83	956
Tajikistan	28.5	105.0	43.2	188.3	166.3	75	730

Source: Narodnoye khozyaystvo SSSR v 1989g., 1990,
    Moscow: Goskomstat.

A   Doctors per 10 000 population
B   Hospital beds per 10 000 population
C   Infant mortality per 1000 births
D   Average monthly pay for industrial workers (rubles)
E   Average monthly pay for collective farmers (rubles)
F   Numbers per 1000 aged 15 and over with completed
      higher education
G   Retail turnover per capita (rubles).

for others. Certain republics can be seen to share many of the features of the less developed 'Third World'; these include all the 'Asian' group. Republics of the 'Intermediate' group, such as Georgia, Armenia and Moldova, seem bent on pushing themselves downwards into this category through inter-republican or civil strife. In contrast, the 'European' Slavic and Baltic republics, for all their difficulties, enjoy vastly better social conditions and though they lag well behind the countries of western Europe, are not entirely out of sight.

References

Narodnoye khozyaystvo SSSR v 1989g. 1990. Moscow, Goskomstat.
Narodnoye khozyaystvo SSSR v 1990g. 1991. Moscow, Goskomstat.
Sidorov D A 1992 Variations in perceived level of prestige of residential areas in the former USSR. *Urban Geography* **13** (4): 355–73.

7

Agriculture and rural development

Judith Pallot

The communist revolution and the countryside

The 1917 revolution did not have an immediate effect upon the nature of rural society in Russia, although in the turbulence of revolutionary upheaval and civil war the peasants were able to seize land from landowners, the church and state to add to their own farm land. The real force of the revolution was felt a decade later when, under Stalin's leadership, the Communist Party launched a campaign to collectivize farming. Collectivization profoundly changed the character of rural society in all parts of the Soviet Union and resulted in the development of a novel system of farm organization consisting of large, vertically structured enterprises (the *kolkhozy* and *sovkhozy*) which were linked into the state's central planning system. The process of creating this structure was traumatic. Hundreds of thousands of independent peasant and nomad households were forced to surrender their land, livestock and farm equipment to the newly formed collective farms. Henceforth, they were to labour collectively with their fellow villagers to produce grain and livestock products for delivery to the state. Despite strong opposition to these changes, which in some regions involved the peasants slaughtering their livestock rather than surrender them to the collectives, the majority of independent farmers and pastoralists had been collectivized by 1933. During the subsequent decades of Communist Party rule some changes were made in the organizational structure of farms and new types of enterprises were introduced into rural USSR, but collective and state farms remained the basic building blocks of the Soviet rural economy.

In theory, collective farms (*kolkhozy*) were producer cooperatives in which all members had equal decision-making rights. In reality, decision making on farms was highly centralized with lines of authority running from the chairman or chairwoman and a board of managers, through 'brigade' leaders, to the ordinary farm workers organized in teams. Under this hierarchical system of command, farm workers were given little responsibility for the work they did. Their daily tasks were determined by those in authority over them, just as, in fact, farm managers were themselves subject to control and direction by state administrative organs. However, farm workers did have scope for independent farming on their supplementary, or 'private', plots. These were small parcels of land (usually under one hectare in size) which collective farms allowed their members for personal use. Initially, these plots were used by the peasants to meet their subsistence needs but, in time, as collective farm wages improved, they were additionally used for commercial production. Collective farm markets, which were found in virtually all Soviet cities and were supplied by 'private' plot production, satisfied a large proportion of the demand for fruit, vegetables, meat and dairy products in the former Soviet Union.

The existence side by side of 'socialized' production on the collectivized fields and 'private' production on the farm workers' personal plots meant that the Soviet farm system had a dual character. But the two sides of farming were strongly inter-dependent. The socialized sector provided inputs into the private sector and the private sector provided a reservoir of labour and produce upon which parent collectives could draw. This dual structure, in turn, was embedded in a complex of state-run institutions and enterprises upon which collective farms depended for the provision and repair of machinery, the supply of their major inputs, the marketing of their output, and for farm building works. Despite a trend in the final years of Soviet power for collective farms to join together in the setting up of inter-collective farm cooperatives in order to overcome supply difficulties, all farms remained strongly dependent upon the state for a whole range of functions.

In addition to collective farms, state farms (*sovkhozy*)

also appeared in the Soviet countryside in the 1930s. State farms were considered to be more 'socialized' than collective farms and superior to them. They were directly owned and managed by the state, their workers were given the same terms of employment as industrial employees and they were, on the whole, larger, more heavily capitalized and more specialized than collectives. At first, state farms were few in number but from the 1950s the balance began to shift in their favour as a result of collective farm mergers and conversions, and the founding of large numbers of state farms in new farming regions, such as in West Siberia and Kazakhstan. On the eve of the Soviet Union's collapse, there were 25 800 collective farms, with an average size of 6 600 hectares, and 21 000 state farms, with an average size of 17 300 hectares in the Soviet Union. By this time, changes in collective farm workers' terms of employment and measures introduced to increase state farms' economic accountability had eliminated the significant differences between the two types of farm. Collective and state farms entered the post-Soviet era more or less on an equal footing and there is no reason to suppose that their post-Soviet futures will be much influenced by their formal status prior to 1991.

Rural society and the problem of rural underdevelopment in the former USSR

One feature of the way in which the Soviet planners approached the countryside was that they tended to view rural and agricultural development as synonymous. Decisions about how to use rural resources, about meeting the social and welfare needs of the rural population and conserving the environment were taken by Soviet planners with an eye to the current needs of agricultural production. For example, in the postwar years the Soviet government became concerned about the low standard of living in rural areas but this concern was prompted by the fact that high levels of rural out-migration were leading to shortages of farm labour. The solutions proposed were correspondingly designed to keep people of working age in agriculture. Poor living conditions *per se* were not viewed as the problem. This can be illustrated by the programme launched by Mikhail Gorbachev in April 1989 ostensibly aimed at improving the conditions of rural life. The preamble to the decree stated that it was necessary to do everything to implement the programme so that it touched 'every rural worker and family' because '. . . ultimately, all this is aimed at achieving sufficient agricultural production in the country and at improving the supply of food for all Soviet people' (quoted in Pallot 1990:661).

Such production-orientated attitudes towards rural development meant that many aspects of rural life and the rural economy in the Soviet Union did not receive the attention they deserved. The potential, for example, of rural areas for recreational development remained largely unexploited, while on the social side the needs of some vulnerable groups, such as the elderly, were overlooked. To some extent the bias in rural planning towards agriculture's needs was understandable; agriculture and its related activities was the major employer in rural areas and collective and state farms physically dominated rural areas. Most people in rural USSR outside district (*rayon*) centres lived within the territory of collective and state farms, whether or not they were employed in agriculture, and they looked to farms for their social, educational and health provision. Furthermore, collective and state farm managers were key figures in local politics and they were able to influence the course of rural development. Thus, although local authorities might have ambitious plans for housing and road building, whether these were executed would often depend upon local farms agreeing to resource the projects. Doctors, teachers and other professionals in many rural districts were paid by farm managements for their services and such recreational facilities as existed in rural USSR were usually organized by collective and state farms as a side-line.

The influential position that farm 'bosses' acquired in rural society and politics in the former USSR was not always used to further the interests of the mass of the rural people. In the years immediately after the collectivization drive, the majority of chairmen and chairwomen of collective farms came from the towns and, charged with the task of fulfilling state orders and 'socializing' the peasants into communist society, they made little attempt to improve the lot of the ordinary farm worker. In later years, the interests of collective farm managements and their farm workers were more likely to coincide but the scope the former had for improving the conditions of rural life was constrained by lack of financial resources and legislation which discriminated against collective farm workers. For example, until the 1970s collective farm workers were not entitled by right to the internal passport which allowed other Soviet citizens to change their place of residence, and there were similar delays in granting them rights to a full state pension and other benefits.

It is possible to identify especially disadvantaged groups within Soviet rural society. The elderly have already been mentioned: they could suffer a double disadvantage from having little pension to show for their working lives spent in agriculture and from the woefully poor provision of rural health services. As one Russian rural geographer observed, countless numbers of pensioners in the Soviet Union lived out their retirement years in rural districts, often in conditions of considerable hardship, at little effective cost to the state (Alekseyev 1988, quoted in Pallot 1990). Women constituted another relatively disadvantaged group of rural workers. Despite optimistic images of women tractor drivers in the 1930s, the average female farm worker has been confined in low skill, predominantly manual, farm jobs and has had little access to positions of authority on farms (Bridger 1987). Like the elderly, women have also suffered from poor health care provision, particularly with respect to the conditions for childbirth. The concentration of female labour in routine and manual tasks has exposed them more than other rural workers to the effects of the chemicalization of Soviet agriculture. This was a particular problem in the Central Asian republics where women and children cotton pickers were forced to suffer the harmful effects of defoliants.

Many of the social problems of the Soviet countryside were associated with the changing demographic structure of the rural population (Table 7.1). The demographic history of the Soviet countryside presents a complex picture. Industrialization and urbanization on the scale experienced in the USSR inevitably led

to a rapid decline in the relative share of the rural population. The decline was greatest in the more developed republics, such as the Baltic states and the Russian republic but it was slower in the Central Asian republics. While these differences were only to be expected, there have been specific features of rural population history in both the heartland of central Russia and the peripheries of Central Asia which have exaggerated 'normal' tendencies. In European Russia rural out-migration in the postwar period developed into a veritable 'flight from the countryside' which has left countless settlements depopulated and rural houses empty. In Central Asia, by contrast, the level of rural out-migration has been less than might have been expected, despite the growth of towns and the development of industry. One reason for this was that opportunities for urban employment were often blocked by in-migrants from other republics who had industrial skills and spoke fluent Russian, the *lingua franca* of the Soviet industrial economy. Even though there was a fall-off in the 1980s in the number of in-migrants to Central Asian cities from other parts of the Soviet Union, the rural population in the republics remained relatively immobile. By the end of the decade underemployment had transformed itself into rural unemployment in the more densely populated agricultural districts in the region such as the Fergana Valley.

It would be wrong to suggest that no attempts were made under Soviet communism to solve the country's rural problems. Compared with the 1930s, rural inhabitants in the former Soviet Union today have better access to hospitals, schools, recreational facilities and retail outlets. Electricity has been taken to most rural regions. Provision has, however, been spatially uneven with some rural inhabitants having better access to services than others. In part, this has come about because of the rural settlement policy pursued by successive postwar Soviet governments. As in other urbanizing countries, the Soviet Union was confronted in its rural regions with the problem of providing services to a declining and dispersed population. The approach it adopted to solving the problem was analogous to the 'key settlement' policy pursued in some Western countries (Pallot 1988). In the Soviet version of the policy all rural settlements were assigned to one of two classes; viable (*perspektivnyye*) or non-viable (*neperspektivnyye*). The former were villages selected for improvement and enlargement while the latter villages were adjudged to have no future and were to be left to 'die out' naturally or were to have their populations relocated. The policy envisaged the progressive concentration of the rural population in an ever declining number of well-serviced

Table 7.1 Rural population in the former Soviet Union

	Percentage of population rural 1990	Rural population change: 1990 as a percentage of 1951	Percentage of rural population pensioners 1989
Russia	26.2	−32.0	22.1
Ukraine	32.7	−28.6	28.3
Belarus'	33.7	−42.9	31.4
Moldova	52.6	+16.2	18.2
Kazakhstan	42.6	+71.7	10.1
Uzbekistan	59.2	+170.0	7.0
Kyrgyzstan	61.9	+114.4	9.3
Tajikistan	67.8	+215.0	6.7
Turkmenistan	54.8	+162.4	6.8
Armenia	31.6	+36.5	13.1
Azerbaijan	46.2	+104.0	9.7
Georgia	44.0	+5.7	19.5
Lithuania	31.5	−33.6	26.3
Latvia	28.8	−25.4	22.9
Estonia	28.4	−19.2	22.5

Source: *The First Book of Demographics for the Republics of the Former Soviet Union*. Maryland, 1992.

rural settlements. It was hoped that this would allow the state to improve rural living standards for a majority of the rural population while minimizing costs of service provision. The aim was to deliver a standard of living to the rural population that was equivalent to urban standards, thereby removing one incentive for people to leave the countryside. In reality, the policy caused considerable misery in rural USSR. Local authorities and farm managements used the classification of settlements as non-viable as an excuse to phase out existing services but, meanwhile, they did not provide sufficient new housing in the viable settlements to accommodate families wanting to move. Even though the viable villages were supposed to be recipients of an urban level of services, shortages of funds meant that plans for improvement were rarely fulfilled. In short, rural settlement policy served to accelerate the demise of large numbers of the Soviet Union's small villages and to increase social differentiation between places. Although the policy was phased out from the late 1970s, its legacy is still felt to this day in the deplorable state of many of the villages of the former Soviet Union. The current crisis of the Russian economy means that the prospects remain remote for upgrading rural services.

One advantage rural inhabitants did have over their urban counterparts was in their housing which often provided them with above-average quantities of living space and access to land for vegetable cultivation. A large portion of rural housing was in private ownership. On the negative side was the fact that many rural houses, whether privately or state owned, lacked basic amenities such as running water and central heating. One course the government might have followed to prevent the deterioration of rural housing, especially in the non-viable villages, was to allow empty dwellings to be developed as second homes for urban inhabitants. Until Gorbachev came to power, restrictions on using rural housing located on farms for anyone other than agricultural or allied workers meant that many properties simply went to ruin. In the 1980s the restrictions were relaxed and measures were introduced positively to encourage reverse migration from the cities but there still remain many unoccupied dwellings in rural regions of the former USSR, especially in European Russia (Fig. 7.1 and 7.2). Under President Boris Yeltsin poor rural services and inadequate communications have continued to deter people from seeking second homes at any distance from the major cities. The principal way the government has been trying to repopulate the deserted villages of northern and central Russia is to settle demobilized soldiers and refugees fleeing from trouble-spots in the non-Russian republics. The countryside is also seen by some as providing a home and 'subsistence' safety net

for the urban unemployed whose numbers seem bound to grow in the future. Whether such disparate groups of settlers will become permanent rural inhabitants is difficult to predict – they certainly will face myriad difficulties adapting to the conditions of rural life.

Environmental problems in the countryside

One of the undesirable legacies of the communist era is the damage that Soviet agriculture did to the environment. In the quest for increased agricultural output, policies were pursued which led to a depletion of the quality of agricultural resources in all regions of the former Soviet Union. The most damaging included the pressure farms came under to extend the ploughed area and the heavy reliance that was placed on chemicals and fertilizers to raise yields. From the 1930s, campaigns to bring hitherto uncultivated land under the plough were an integral part of Soviet agricultural policy and they reached their apotheosis in Khrushchev's Virgin Lands campaign launched in the second half of the 1950s. This campaign added to Soviet ploughland by approximately 50 per cent but at the cost of creating dust bowls in the more marginal regions in Kazakhstan and West Siberia. At the local level the extension of arable at the expense of other land uses took place on nearly all farms in the Soviet Union, while in the arid areas poorly executed irrigation schemes added to the Soviet Union's wasted agricultural resources as a result of secondary salinization. Poorly adapted rotations, vertical ploughing, the use of heavy machinery which compacted the soil, and continuous cropping are just some of the farming practices that have exacerbated the problem of deteriorating soil conditions in all regions of the former USSR.

It is difficult to be precise about the scale of damage done to the former Soviet Union's agricultural resources. One authority has estimated that in 1989 nearly half of all the farm land in the USSR was 'seriously imperiled' by such processes as salinization, erosion and waterlogging (Feshbach and Friendly 1992:57–8). Over the last twenty years the levels of humus in the soil have dropped by between 8 and 30 per cent, ravines and gullies have 'consumed' 6.5 million hectares of agricultural land and 14 per cent of all irrigated land has required restorative work because of waterlogging and salinization. The most damaged regions are in the North Caucasus and central black earth regions of the Russian Federation where some 12 per cent of all arable land loses more than ten tonnes per hectare of topsoil annually and a further 12 per cent loses between five and ten

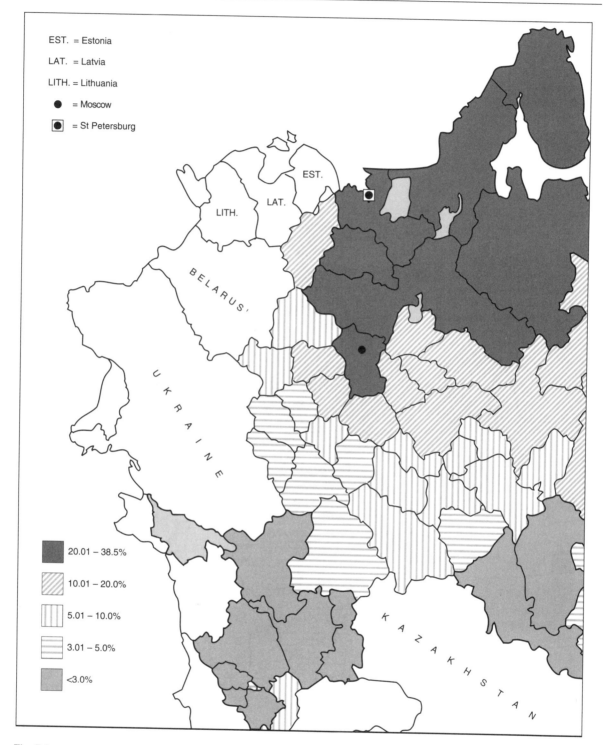

Fig. 7.1
Percentage of rural houses unoccupied in 1990 in European Russia.

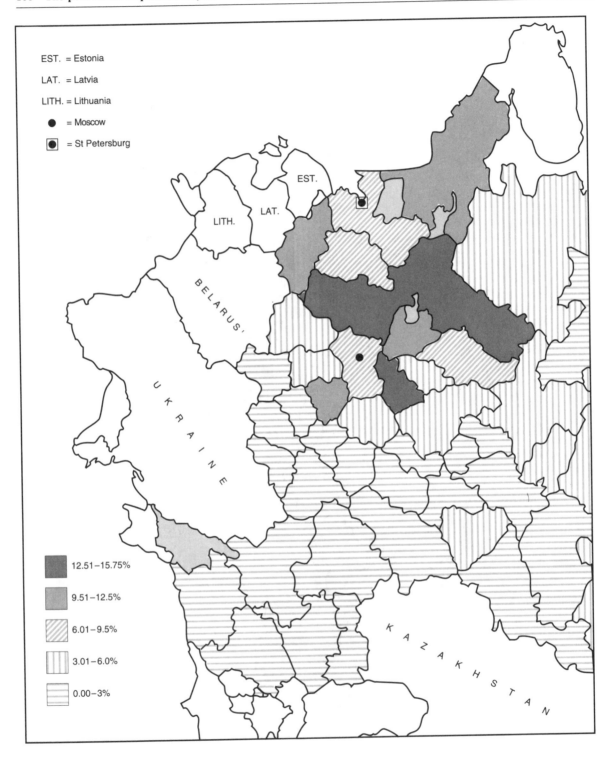

Fig. 7.2
Percentage of rural housing in European Russia used as second homes, 1990.

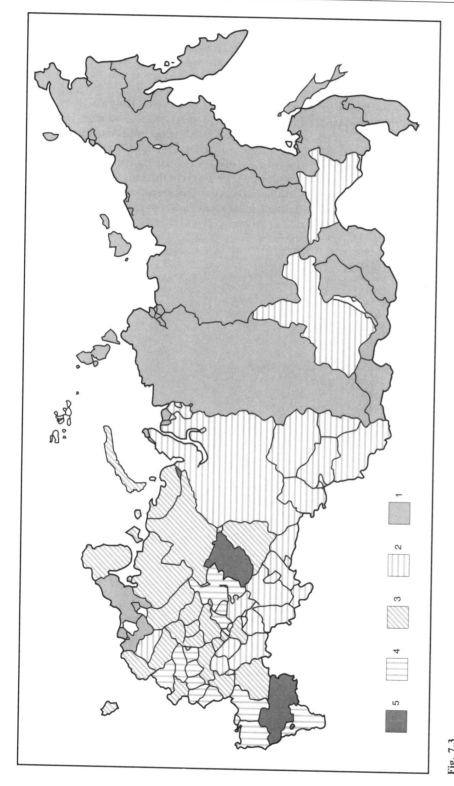

Fig. 7.3
The degree of degradation of arable land and pastures in the Russian Federation on a scale from 5 (most degraded) to 1 (least degraded). (Based on a complex index including measures of water and wind erosion, soil salination and soil compaction.)
Source: T. G. Nefedova, *Sel'skoe Khozyaistvo Rossii Nakanune Reform 1990–1992* (Moskow 1992)

tonnes per hectare. In the Volga region and Ukraine, where on average soil humus content has declined by 9 per cent in the past twenty years, the deterioration of the soil is also a major problem. All are major regions of cereal cultivation. Pastures have also been afflicted by a deterioration in underlying soils, the worst affected areas for this problem being Dagestan and other livestock regions in the North Caucasus and Kazakhstan (Fig. 7.3).

The use of chemicals on the land was a feature of Soviet agriculture in the decades after 1960 when they were believed to hold the key to overcoming the low productivity of Soviet agriculture. Much of the fertilizer and pesticides destined for agriculture suffered from the normal bottlenecks of the Soviet system. Storage facilities and the machinery needed to apply the chemicals were often inadequate and the effectiveness of the chemicals was thus reduced. Furthermore, agrochemicals constituted a serious health hazard for the Soviet population as safety standards were frequently flouted. Ministry of Health reports of the late 1980s showed that levels of nitrates in certain foods exceeded safe levels, while the reversal of a ban on the use of DDT in 1970 led to a number of cases of soil and water body contamination (Feshbach and Friendly 1992:65). In the wake of Chernobyl', the possible radioactive contamination of food was another health hazard faced by Soviet consumers.

Table 7.2(a) Selected countries of the former Soviet Union in global comparison

	Countries	Grain yields (centners*/ha)	Mineral fertilizers (kg/ha)	Density of roads (km/km²)	Density of population (head/km²)
Group 1	USA	43.4	95	387	27
	Japan	56.6	425	2005	323
	Austria	54.1	210	1295	92
	Belgium	62.3	505	4073	326
	W. Germany	57.1	405	2013	251
	UK	57.5	359	1467	237
Group II	Bulgaria	39.9	199	305	82
	Hungary	47.7	258	569	114
	Poland	31.4	224	728	126
	Romania	26.7	135	160	101
	Portugal	16.7	75	488	112
Group III	Canada	22.0	47	27	3
	Argentina	22.6	95	21	12
	Philippines	19.4	64	75	209
	Thailand	20.8	33	76	109
	Turkey	20.6	62	59	73
Group IV	Mongolia	12.2	15	1	1.4
	Afghanistan	13.4	33	4	25
	Iraq	9.0	42	88	43
	Cameroon	12.3	–	7	25
	Gambia	9.4	3	5	23
Group V	Estonia	21.1	283	65	35
	Lithuania	26.7	298	54	57
	Belarus'	25.2	339	32	49
	Ukraine	30.6	152	39	86
	Moldova	33.9	192	41	130
	Kazakhstan	10.0	33	4	6

Group I	High productivity agricultural economies
Group II	Medium productivity agricultural economies of the East European countries
Group III	Medium productivity agricultural economies of the Americas and Asia
Group IV	Low productivity agricultural economies of Asia and Africa
Group V	Former Soviet republics.

* 1 centner = 50 kg

Agricultural performance in the former Soviet Union

Despite large investments in agriculture in the postwar period, Soviet agriculture consistently failed to meet its production targets. By the 1970s it had become a drain on the Soviet economy – while absorbing more than 25 per cent of annual investment, agriculture suffered a declining rate of growth. Furthermore, in order to sustain stable prices for the consumer, the state found itself having to make ever greater subsidies to the agricultural sector. By the end of the Soviet period the direct subsidies were running in the region of US$50 billion annually. Soviet agriculture had turned out to be costly both to the budget and in terms of forgone opportunities to produce and consume other goods and services. The clearest evidence of Soviet agriculture's failure was that despite abundant agricultural resources, the state had to import cereals and other foodstuffs from the West.

Tables 7.2(a) and 7.2(b) use various indices to show how agricultural production in the former Soviet Union compared with other countries of the world in the period just preceding its dissolution. They show the extent to which production in all parts of the USSR had fallen behind the developed world. Nowhere in the Russian republic, for example, were there any regions that could be compared with the high yielding agrarian economies of western Europe. However, yields in a small number of regions (North Caucasus, Belgorod *Oblast* and Moscow *Oblast*) could be compared with east European countries. These are regions with good natural conditions for farming, adequate infrastructure and dense population. The central black earth oblasts, which also have favourable environmental conditions, and Leningrad Oblast recorded yields similar to those in the Americas but by far the majority of oblasts in the Russian republic recorded yields no better than Third World countries. The non-black earth centre with yields of fifteen centners (750 kg) per hectare was equivalent only to the poorest Third World countries. Outside the Russian Federation, Ukraine, Moldova, Lithuania and Estonia could be said to be equivalent

Table 7.2(b) Selected oblasts in the Russian Federation: Rural economic indicators

Oblast	Grain yields (centners*/ha)	Mineral fertilizers (kg/ha)	Density of roads (km/km²)	Density of population (head/km²)
Krasnodar	38.4	195	28	62
Kabardino-Balkar'	34.3	164	19	62
Moscow	25.9	226	58	334
Belgorod	25.8	167	26	52
Stavropol	24.6	109	11	36
Kursk	23.3	233	37	45
Voronezh	22.2	133	23	47
Orel	18.4	198	24	36
Leningrad	16.3	240	32	78
Bryansk	14.6	230	31	42
Ivanov	13.9	183	14	55
Nizhegorod	14.4	174	18	50
Samara	13.3	51	26	61
Novosibirsk	12.6	27	9	16
Khabarovsk	13.5	152	3	2
Komi	9.5	374	4	3
Pskov	8.7	176	21	14
Perm	8.9	122	9	19
Amur	9.2	111	6	3
Tuva	9.1	30	1	2

Source: T.G. Nefedova, *Sel'skoye khozyaystvo rossii nakanune reform 1990–92* (Moscow, 1992).

* 1 centner = 50 kg

to the east European group of nations, and Kazakhstan to the Asiatic and African group. In Kazakhstan an important factor explaining low yields is the marginal nature of the environment which is susceptible to droughts. A characteristic feature of harvests in this region is their instability.

The performance of Soviet farms was uneven geographically. Yields declined along a south–north gradient in European USSR and along a west–east gradient from European USSR to Siberia and north Kazakhstan. Research by Soviet scientists showed that only part of this variation could be explained by differences in bio-climatic resources. For example, in most oblasts including those with uniform natural conditions, there were steep gradients in yields, of the order of 200 to 300 per cent, between sub-urban and peripheral zones. There were also large gradients in yields between oblasts lying either side of the political boundaries between the Russian, Ukrainian and Baltic republics. In few Russian oblasts did productivity compare well with Ukraine and the Baltic states. The central black earth and Volga oblasts which were major cereal producing regions and non-black earth oblasts all were producing well below their potential. This is further evidence that socio-economic, rather than environmental, factors were at fault. Even when the Soviet government made attempts to increase agricultural performance in the Russian republic, the results could be disappointing. The non-black earth belt was a case in point. This region was the recipient from the 1960s of preferential allocation of resources for intensification. Despite increased investments in fertilizers, machinery and infrastructure, there was a decline in yields in the region in the 1970s. The element that had been overlooked by planners was the availability of labour. Out-migration in the postwar years had, in fact, left the region without the labour force able to make use of the new inputs.

Spatial inefficiency in the Soviet agricultural economy

Much has been written about the causes of Soviet agriculture's failure. Analyses divide between those that blame the structure of collective and state farms and those that locate the problems more in the command–administrative planning system. The charge against Soviet farm structure was that farms were too large and their management too centralized for efficient farming. Ordinary farm workers were given inadequate material incentives and too little responsibility, with the result that they were not encouraged to improve productivity, reduce costs or preserve capital and land. According to this argument, the fundamental problem was that the farm workers did not own the land.

Current attempts in many of the Soviet successor states to divide collective and state farms into smaller units and to privatize the land stem from this analysis of the agricultural problem. The other explanation, which puts the blame for agriculture's failure on the command–administrative system, drew attention to the inefficiency of using procurement quotas (targets) and administratively fixed prices, rather than the free market, to plan agriculture. Because prices and costs had no economic meaning in the Soviet economy, the command structure resulted in misallocation, waste and the over-use of inputs. One particular example of such inefficiency was in the rather weak specialization of many farm units.

Under the Soviet system of price fixing, the prices that the state paid for agricultural commodities were set in relation to the costs of production. While the system in theory enabled all farms to cover their costs, it tended to suppress specialization and led to various anomalies. One problem was that livestock products were generally less well rewarded than cereals and as a result farms that otherwise would have had a comparative advantage in meat production were reluctant to develop this branch of husbandry. Local planners responded by 'spreading the burden' of the livestock procurement quota between all the farms in a region. Such a practice of equally distributing procurement quotas among farms was applied to other products as well and it was the principal reason why the majority of collective and state farms in the Soviet Union were mixed units with a diversified production profile.

A similar failure for specialization to develop afflicted a higher spatial scale because of the existence of regional price zones. Differentials in the price of wheat were zoned in order to cover the costs of production in different geographical regions but this meant that collective farms producing high cost wheat in eastern regions of the USSR were just as well rewarded as farms producing it more efficiently in the North Caucasus and Ukraine. Even in sub-urban zones the degree to which collective and state farms specialized in market garden and dairy produce was less than might otherwise have been expected had production taken place in response to the market.

Lack of specialization at the regional and sub-regional level was combined in the Soviet Union with over-specialization at the republic level. This also came about because of planning decisions. The central planners, rightly, realized that the Baltic republics had a comparative advantage for intensive livestock production and dairying and so abolished cereal quotas for the republics. In the case of the Central Asian republics, a similar policy, which was designed to enable the Soviet Union to be self-sufficient in cotton, was taken too far. The high cotton procurement targets demanded from republics such as Uzbekistan meant that basic foodstuffs, which could have been produced locally at low cost, had to be imported from other parts of the Soviet Union into the region. Since the break-up of the Soviet Union, market prices have begun to replace administratively fixed prices for agricultural commodities and in the Russian Federation regional price zones for cereals have been abolished. In the long term these changes should result in a realignment of patterns of production at regional and farm level in all the republics.

The impact of the Soviet Union's disintegration on food production in the newly independent republics

The long-term aim of all the successor states to the USSR is to restructure their agrarian economies in order to overcome the inefficiencies of the former system. However, in the short term the majority have had to adopt a variety of 'survival' strategies to overcome the disruption caused by the collapse of the USSR. The collapse had two immediate negative impacts on agriculture and the food economy in the former republics: firstly, all experienced a decline in agricultural output (for example, in the Russian Federation agricultural output was only 65.8 per cent of its 1990 level in 1993) which was associated with general economic crisis, and secondly, all suffered from disruption in normal inter-republican trade in food and agricultural inputs. These problems have been exacerbated in the case of Georgia, Tajikistan, Armenia and Azerbaijan by ethnic conflict and civil war which have severed trade routes and diverted resources away from agriculture. The successor states all share the challenge of securing food for their populations.

In the past, agricultural shortfalls were met by imports from the West. The decline in output experienced by the successor states has meant that the dependence on imports has not abated. However, as the successor states all face financial constraints it is likely that food imports will be at a lower level than in the past and to a greater or lesser extent all the successor states will rely upon food aid. Each state now handles its own foreign food imports and is responsible for negotiating export credit guarantees and food aid packages with foreign powers.

Trade between the successor states was heavily disrupted by the break-up of the USSR. The immediate reaction of agricultural producers to the economic and political instability of 1990–91 was to withhold agricultural produce from the market and this was often encouraged by regional authorities worried about local food security for the population. Such actions led to all the successor governments having to cope with shortfalls in procurements and this affected their ability to supply their traditional customers in other republics. Most also began to search for new customers who would pay for agricultural commodities in hard currency or in shortage goods.

Table 7.3 shows how the food trade between the former Soviet republics stood at the end of the 1980s. Clearly the republics set out on the path of independence with differing abilities to feed themselves. In potentially the strongest position were Ukraine and Moldova, both of which had positive food trade balances with the other republics. Ukraine supplied Russia, Central Asia and the Transcaucasus with grain, vegetable oil, meat and dairy products, while Moldova sold half of its total output of fruit and vegetables to the rest of the USSR but was dependent upon grain imports. Since 1990 the favourable position of Ukraine has been eroded. Long considered the 'bread basket' of the USSR, Ukraine has been

Table 7.3 Net imports (excluding external trade) of primary foods by republic of the former Soviet Union 1988–90

Republic	Grains	Meat and meat products	Milk and dairy products	Sugar	Vegetable oil	Fruit	Vegetables
Russia	20 383	1 663	8 598	2 409	495	1 493	1 613
Ukraine	−848	−361	−1 856	−3 570	−262	−96	−269
Belarus'	3 342	−265	−1 887	161	60	53	137
Moldova	686	−63	−92	−157	−65	−577	−205
Kazakhstan	−3 350	−153	202	236	26	10	−33
Baltic states	3 718	−84	−686	−11	13	16	4
Transcaucasus	4 219	72	1 118	201	15	−162	−83
Central Asia	7 035	70	506	115	−50	−98	−228

Source: *Strany - chleny SNG: statisticheskiy yezhegodnik*, 1992. Statkom SNG

experiencing trouble supplying its own, let alone others', food needs. Grain exports have dwindled since 1990 and may even have turned to imports in 1993. A major reason for the continuing crisis in Ukrainian agriculture is the loss of inputs such as fertilizers and machinery which in the past were imported from other republics. Unlike in Russia, yields have been slow to recover in Ukraine from the lows that immediately followed the Soviet Union's disintegration.

By contrast with Ukraine, the Russian Federation was always a net food importer and has relied on grain imports to supply its deficit regions in the north, Siberia and the Far East. Russia was also a large importer of meat and dairy products which were used, in particular, to supply the large urban markets of Leningrad (St Petersburg) and Moscow, and it was a major purchaser of Ukrainian sugar. There has been no decline in Russia's import dependency since 1990 but, increasingly, surplus demand has been satisfied by countries outside the former Soviet Union rather than by Russia's immediate neighbours. It is likely that the Russian Federation will need to import some 12 per cent of its total cereal demand for the foreseeable future.

Kazakhstan was a major, although irregular, supplier of grain to other Soviet republics. Since the break-up of the USSR it has continued to supply Russia, Belarus', Uzbekistan and Turkmenistan with wheat, except in 1991 when a disastrous harvest meant it had to suspend trade. Kazakhstan remains, as before 1990, dependent upon imports for meat, dairy products and vegetable oil. The Transcaucasian and Central Asian states are in the most vulnerable position with respect to their ability to feed themselves because of their heavy dependency upon cereal imports (Fig. 7.4). Georgia, Armenia, Azerbaijan and Tajikistan have an acute need for food which is being partially met by humanitarian aid.

It is inevitable that in the course of time the Soviet successor states will attempt to restructure their agricultural profiles in order to overcome the problems they are currently facing. Although in the short term each state is pursuing policies aimed at increasing self-sufficiency in food, in the longer term it can be assumed that under the influence of the market, the comparative advantages and disadvantages each has will play an increasing role in shaping production. Unfortunately, current food insecurity in some places

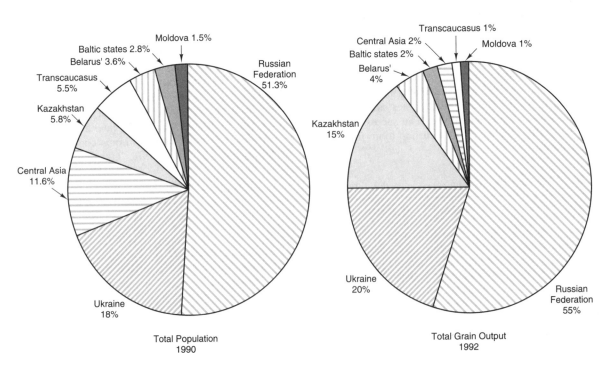

Fig. 7.4
A comparison of the percentage share of the total population and grain output of the republics of the former Soviet Union.

is undermining the potential for future recovery. In Georgia, for example, vineyards have been grubbed up or left to run wild, which can only damage the prospects of recovery in the formerly flourishing wine industry.

Although all the successor states are committed to introducing market elements into the agricultural sector, the market environment was still relatively weak at the end of 1993. In the Russian Federation marketization has progressed further than in the other former republics, with the exception of the Baltic states. During the last months of 1993, a number of presidential decrees were passed which abolished food subsidies to consumers, ended state procurements in agriculture and privatized the food processing industry. The state will continue in Russia to have a role in supplying grain to deficit regions and the large urban centres but since the end of 1993 has been bound to pay market prices for any purchases it makes.

Summarizing the impact of the Soviet Union's dissolution on agriculture in the former republics, three main trends can be observed:

1. The attempt on the part of each of the successor states to restructure their agricultural sector in order to reduce dependency upon food imports from other successor states. This has been reflected in increases in the sown area put under grains in Central Asian states, Georgia and Armenia (Table 7.4). The president of Turkmenistan, for example, has resolved to make his state self-sufficient in grain by the year 1995.

2. A reduction in the total sown area of all but a few of the Soviet successor states. This has been particularly marked in those regions of the former Soviet Union where high prices paid for cereals had resulted in the expansion of cultivation into marginal areas. Reductions in cereal acreages have taken place in the Russian Federation, Ukraine, Kazakhstan, Latvia, Tajikistan and Georgia. In the latter two this is, no doubt, associated with the disruption to normal agriculture caused by internal conflict.

3. A universal decline in livestock husbandry relative to crop production. This is a consequence partly of a fall-off in demand as subsidies to the consumer for meat and dairy products have been reduced, and partly of shortages of feed resulting in shrinkages of herds. It will take some years for livestock numbers to build up again but some Western analysts have observed that levels of meat and dairy consumption were, in any case, too high in the former Soviet Union given the per capita income of the country. It might be, therefore,

that the decline in meat production in the former Soviet Union represents consumption settling to its 'proper' level and that for this reason herds are unlikely to expand in the near future, even if there is an economic upturn.

Land reform as the long-term solution to the successor states' agricultural problems

All the successor states to the USSR have declared that the long-term solution to their agricultural problems lies in land reform involving the privatization of collective and state farms and the promotion of independent, or 'peasant' farming (Fig. 7.5 and Table 7.5). To date, the path of land reform has been far from smooth. It has proceeded furthest in the Baltic states, Armenia and the Russian Federation but elsewhere it has made modest headway. In Kazakhstan and Turkmenistan ethnicity has interfered with plans to privatize farming. In the Kazakh case prospects of land passing into the ownership of ethnic Russians has brought land reform to a halt in the northern oblasts of the state and similar fears of land passing out of the control of the indigenous population have stopped land reform in the Chu valley in Central Asia. At the opposite extreme, Armenia has largely decollectivized. In Armenia, land of collective and state farms has been divided between peasants into smallholdings with an average size of two hectares.

Table 7.4 Sown area of all grains 1987 and 1992 by republic of the former Soviet Union

	1987	1992	Percentage change
Russia	61 456	57 100	−7.1
Ukraine	13 522	12 100	−10.5
Belarus'	2 442	2 520	+3.2
Moldova	564	733	+30.0
Kazakhstan	24 083	21 960	−8.9
Uzbekistan	992	1 180	+19.0
Kyrgyzstan	544	561	+3.1
Tajikistan	221	215	−2.7
Turkmenistan	187	352	+88.2
Armenia	129	141	+9.3
Azerbaijan	450	572	+27.0
Georgia	269	257	−4.5
Lithuania	972	1 055	+8.5
Latvia	653	650	−0.5
Estonia	357	400	+12.0

Source: Strany – chleny SNG: statisticheskiy yezhegodnik, 1992. Statkom SNG.

		Total number of peasant farms		
> 15 000		5000–10 000		< 2000
10 000–15 000		2000–5000		

(Totals shown for oblasts and republics of the Russian Federation, and eleven successor states)

0 ———— 1000
km

In the Russian Federation land reform has followed a tortuous path and, even though the initial aim of the government was to reinstate peasant farming in the country, current changes in the organization of agriculture promise to create a differentiated farm structure (Van Atta 1993). Under laws and decrees issued during the past few years, collective and state farms in the Russian Federation have been required to set up internal land reform committees to decide upon their future. Approximately half the

EAST SIBERIAN SEA

BERING SEA

SEA OF OKHOTSK

SEA OF JAPAN

- - - - Oblast borders within
the Russian Federation

Fig. 7.5
Spatial distribution of independent 'peasant' farms in the
former Soviet Union, mid-1993.

farms in the country opted to stay as collective or
state farms but the remainder decided to reorganize
themselves along different lines. Some, for example,
have reconstituted themselves as joint-stock companies
issuing shares to their membership or to outsiders,
others have decided to reconstitute themselves as
peasant farming cooperatives, and yet others simply to
divide into separate independent farms. Commentators
observed that in many cases the changes taking place
on farms in the wake of these 'reorganizations'

Table 7.5 Private (peasant) farms in selected republics*

	Number of private (peasant) farms	Average size (in hectares)	Area as a share of total arable (in %)
Russia	261 400	42	6.0
Ukraine	14 400	20	0.9
Belarus'	2 000	19	0.6
Moldova	500	3	0.1
Kazakhstan	8 500	412	9.9
Uzbekistan	5 900	8	1.0
Kyrgyzstan	8 600	44	26.8
Tajikistan	40	25	0.0
Turkmenistan	100	11	0.1
Armenia	243 000	2	81.8
Azerbaijan	200	39	0.5

*Figures for Russia 1 August 1993; figures for other republics 1 January 1993.

were little more than cosmetic. It was in order to give some real force to the land reform, that Yeltsin issued decrees at the end of 1993 aimed at strengthening individuals' rights *vis-à-vis* their parent farms. These included giving legal recognition to group or cooperative farms, accelerating the process of issuing legal documents of land ownership to individual peasant farmers, and making parent farms legally obliged to honour shares issued to their members under the government's voucher programme in land or in cash. Most importantly, at the end of 1993 the Yeltsin government removed restrictions that had previously prevented a free agricultural land market developing. There is no question but that these measures were designed to deliver a death blow to the country's remaining state and collective farms.

The formation of individual farms has been taking place steadily in Russia since the final years of Gorbachev's presidency of the USSR. In mid-1993 their number stood at 261 393. These farms had been formed in a variety of ways. Some were fashioned out of the land of subdivided collective and state farms, others were formed by non-agricultural workers on land belonging to local authorities, and yet others were formed by collective farm members unilaterally withdrawing from their parent farms. The size of these farms varies regionally, but with an average of 42 hectares, Russia's new private farms can be described as small. However, their owners constitute a diverse socio-economic group embracing pensioners, the 'new entrepreneurs', peasant farmers, refugees and

urbanites. At one end of the scale, there are some private farms that are clearly nascent capitalist units and at the other, there are economically marginal farms which are under-resourced and lack the potential to provide their owners with a rising standard of living. Because of the variety of situations they embrace, it is important to interpret government statistics about the creation of independent farms with caution.

Notwithstanding Yeltsin's 1993 measures to accelerate the land reform, it is likely that the formation of private farms in the Russian Federation will continue to be beset by problems. It is unpopular among both farm managements and farm workers. The managers' opposition to the land reform is understandable because acceptance of it will lead to their losing power and privileges. On the farm workers' part it is the risks and uncertainties associated with setting up on independent farms that have tended to make them cautious in taking up their new rights. Lack of access to affordable credit, the unavailability of appropriate technologies and undeveloped marketing structures could hamper the reform in the future, as they have in the past. Furthermore, before farm workers can be encouraged to leave the security of their collective farms, they will have to be confident that the local authorities will step in to provide an adequate level of services needed to meet their demand for education, welfare and consumption.

There can be no doubt that the penetration of market forces into the countryside through land privatization and price reforms will gradually erode resistance to change in the post-Soviet countryside but it will be some years yet before the last of the old collective and state farm structures are seen. For the foreseeable future it is most likely that large and medium sized farms with forms of joint ownership and production will continue to coexist alongside private 'capitalist' farms and marginal peasant farms in the post-Soviet countryside.

References

Bridger S 1987 *Women in the Soviet Countryside*. Cambridge University Press, Cambridge.

Feshbach M, Friendly A, jr 1992 *Ecocide in the USSR* Aurum Press, London.

Pallot J 1988 The USSR. In Cloke P (ed.) *Policies and Plans for Rural People: An international perspective*. Allen and Unwin, London.

Pallot J 1990 Rural depopulation and the restoration of the Russian village under Gorbachev. *Soviet Studies* **42** (4): 655–74

Van Atta Don 1993 Yeltsin decree finally ends 'Second Serfdom' in *Russia. Radio Free Europe/Radio Liberty Research Report* **2** (46) November: 33–9.

8

Urban development

Denis J.B. Shaw

One of the most significant processes to affect Soviet society in the period between 1917 and the end of the 1980s was its rapid urbanization. Soviet power witnessed the transformation of what had previously been a largely rural and peasant country into a highly urbanized and industrialized one. Thus, whereas only about 17 per cent of Russia's population had lived in towns and cities on the eve of the 1917 revolution, almost two-thirds of the Soviet population did so by the time the USSR ceased to exist in 1991. Urbanism had therefore become part and parcel of the life of most Soviet citizens. Yet this urban heritage was not of equal weight in all the republics (Table 8.1). Whereas the most developed republics, namely the Baltic states and the Slavic republics, were more urbanized than the USSR on average, the southern ones were less so, with the single exception of Armenia. The four republics of Central Asia had less than half of their populations living in urban places, and in the case of Tajikistan the urbanized proportion was still less than one-third. Even so cities and towns are central to the economic, political and social well-being of all the successor states without exception. Thus, in seeking to understand the character of the cities and the forces which are now changing them, this chapter focuses upon phenomena which are of vital importance for the future of the republics. Before we can discuss their present-day character, however, we must know what the cities have inherited from the past.

Urbanization in the Soviet period

It would have been very surprising if the Bolsheviks who seized power in 1917 had not had strong opinions about Russian towns and cities. After all, they had seized power in the name of the urban working class, many of whom were living in abject circumstances in the handful of industrial towns and cities which then existed. To improve the material conditions of life of

their chief supporters was therefore an important goal for the Bolsheviks. But that was not all. As we have seen, the Bolsheviks were modernizers who believed that industrialization and urbanization were vital to the building of a socialist society. The reality which faced them was a mainly peasant society, most of whose members were suspicious of or even deeply hostile to the Bolsheviks and largely indifferent to socialism. Outside the few industrial cities and regions in European Russia, society was mainly rural and chiefly engaged in traditional forms of agriculture, with the scattered towns having administrative, trading and service functions rather than manufacturing. It was unpromising soil for the cultivation of socialism.

Table 8.1 The fifteen post-Soviet republics: degrees of urbanization at the end of the Soviet period

	Total population 1991 (millions)	Percentage urban
USSR	290.1	66
Russia	148.5	74
Ukraine	51.9	68
Belarus'	10.3	67
Uzbekistan	20.7	40
Kazakhstan	16.8	58
Georgia	5.5	56
Azerbaijan	7.1	54
Lithuania	3.7	69
Moldova	4.4	48
Latvia	2.7	71
Kyrgyzstan	4.4	38
Tajikistan	5.4	31
Armenia	3.4	68
Turkmenistan	3.7	45
Estonia	1.6	72

Source: *Narodnoye khozyaystvo SSSR v 1990g.* 1991.
Finansy i statistika, Moscow: 68–73.

During the first few years of their rule, the Bolsheviks were barely in a position to do much about the cities. Their early policies included the nationalization of major urban industries, land and important properties, and they were able to benefit their proletarian supporters by redistributing housing confiscated from the aristocratic and bourgeois classes. But the outbreak of civil war in 1918 set things back and the accompanying violence and political and economic turbulence led to mass migrations into the countryside as people fled to the villages in search of food. It was only as things slowly recovered during the 1920s that the migrants came back and urban life returned to normal, encouraged among other things by Bolshevik permissiveness towards private business (the 'New Economic Policy' period).

In the late 1920s, however, things changed abruptly. Stalin's industrialization drive inevitably meant urbanization on a massive scale (Table 8.2; Fig. 8.1). Existing industrial cities expanded enormously. In addition, many non-industrial towns and settlements were transformed into industrial ones and very many new towns and cities were created. In the quest to make the USSR as self-sufficient as possible, urbanization and industrialization spread into remote regions as new resources were discovered and exploited, often with the help of forced labour. Urbanization was thus widespread, but by no means even. As noted in Chapter 4, Stalin's policies had the effect of reinforcing the economic dominance of the European core, a dominance inherited from tsarist times. And although numerous cities and towns now developed in the periphery, many regions both in the European territory and beyond remained less urbanized, having relatively little to contribute to Stalin's goals. Moreover, his collectivization policies in the countryside only served to reinforce social differences between rural and urban areas.

Stalin's policies transformed the life of the cities in almost every way (Fig. 8.2). Politically, he finalized the process of centralization which had begun soon after the 1917 revolution. City governments (the city soviets and their executive committees) had almost no autonomy and were subjected to the rule of the party, and more particularly to that of Stalin's agents in the secret police and elsewhere. Dissent or even independent thought became almost impossible in an atmosphere of indiscriminate terror. Since so much building in cities and so many decisions affecting cities were now effected by the centralized industrial ministries or their local enterprises, it became virtually impossible for the city soviets to plan their cities in any co-ordinated way. In fact, in a sense, urbanization under Stalin became almost spontaneous, an ironical

Table 8.2 Urbanization in the USSR, 1913–91

Year	Total population (millions)	Total urban population (millions)	Percentage urban
1913	159.2	28.5	17.9
1940	194.1	63.1	32.5
1959	208.8	100.0	47.9
1970	241.7	136.0	56.3
1979	262.4	163.6	62.3
1989	286.7	188.8	65.9
1991	290.1	191.7	66.1

Sources: Narodnoye khozyaystvo SSSR za 70 let 1987. Finansy i statistika, Moscow: 374–5; *Narodnoye khozyaystvo SSSR v 1990g.* 1991. Finansy i statistika, Moscow p. 67.

situation in what was meant to be a centrally planned system.

Economically, Stalin's policies meant that most investment was ploughed into industry to the detriment of housing, services and other vital parts of the urban infrastructure. The massive immigration into the cities from the countryside could only mean unprecedented overcrowding in urban housing, a feature much exacerbated by the bombings and urban warfare of the Second World War (1941–45) period. Attempts made in the early 1930s to control the growth of the biggest cities by introducing a system of residence permits and by trying to curb the construction of new enterprises there proved only partially effective. In any case, such a response was essentially negative. What was required was a huge effort to build housing and associated services, and this is precisely what Stalin refused to sanction. The new housing which was built under Stalin was chiefly meant for privileged functionaries of party or state or created by important ministries for their own workers. The slightly greater effort made in the postwar period had little effect on the shortages. It was Stalin, moreover, who collectivized the peasantry, thus eliminating peasant trade in foodstuffs and other goods (with the exception of the restricted *kolkhoz* markets), and who abolished the remaining forms of private enterprise. Unfortunately, the state planning of agriculture, retail trade and services proved unequal to the task of catering adequately to the needs of the cities.

Socially, Stalin's policies completed many processes which had been set in train by the Bolshevik revolution while starting others. Thus the remnants of the bourgeoisie and the petty traders and middlemen of the 1920s disappeared. The generation of specialists and officials trained under tsarist conditions was now replaced by Soviet-trained equivalents. Mass migrations into the cities greatly increased the numerical importance of the industrial working class, from whose ranks many were now recruited to join the educated officials and specialists of the Soviet 'intelligentsia'. At the same time, the sheer influx of peasants may have served to preserve a 'rural' character in cities, at least for a time, and inhibited a full adjustment by migrants to the urban environment. Peasant migrants, in other words, may have hindered the appearance of a fully modernized, urbanized society of the kind anticipated by the communists (Lewin 1988). This once again suggests that Stalin's control over the urbanization process was only partial, despite his regime so often being described as 'totalitarian' (Schapiro 1972). One further social development of the Stalin period is also of significance. In spite of the official Soviet goal of social equality, Stalin reintroduced the idea of additional rewards going to key workers and to important officials of party and state. In this way a new form of social inequality threatened to replace that which had been abolished with the tsarist regime.

Many of Stalin's policies were modified by his successors but the essential framework he had created endured to the very end of the Soviet period. Perhaps the key change to affect the cities was the decision in the mid-1950s greatly to increase the construction of housing and services and generally to raise living standards (see Ch. 6). During the following 30 years Soviet cities were enveloped by huge housing projects, largely in the form of multi-storey apartment blocks, as the suburbs sprawled into the surrounding countryside. Investment also went into other aspects of the urban infrastructure, and an attempt was made to place city planning on a sounder basis by giving more power and authority to the city soviets. Yet, despite these major changes, the economy's industrial bias remained unaffected, and insufficient investment was available for housing and other needs. Moreover, given the entrenched character of the Soviet bureaucracy and political system, there was little chance that city planning would succeed in the face of ministerial and industrial interests.

Soviet urban society and associated urban problems

Despite the elements of inequality which began to enter Soviet society from Stalin's time onwards, the Soviet state remained outwardly committed to the ultimate goal of full social equality. Peripheral republics, regions and cities certainly benefited from transfers of capital from the more advanced areas.

Within cities, the commitment to equality was reflected in such policies as normative controls over housing development, low-fare public transport, low-rental housing provision, and state-funded welfare. As time went by, however, the social inequalities inherited from the past did not seem to be disappearing, as they should have done if socialism were to advance. Indeed, in certain respects, inequalities were growing. At the inter-urban level, for example, cities at the top of the urban hierarchy enjoyed better living standards than those lower down, as noted in Chapter 6. Since the economically and politically important cities were the places where the Soviet élite generally lived, the urban hierarchy appeared to reflect an emerging social pyramid. Moreover, the fact that such places were protected in their privileges by the system of internal passports and residence permits reinforced an impression of structural inequality (Zaslavsky 1982). Such cities were only able to survive economically by attracting commuters from less privileged settlements in the surrounding region to do unremunerative jobs (a case of the urban exploitation of the suburbs, the opposite situation from that often found in the United States). A similar role was performed by people on limited-life residence permits, the so-called *limitchiki*. In fact, had it not been for the residence controls, big Soviet cities would certainly have been bigger than they actually were. This fact has given birth to the concept of 'underurbanization', which is said to be a typical feature of communist countries (Ofer 1976). This situation is thus the opposite of the 'overurbanization' which is claimed as a characteristic of many countries in the Third World.

At a broader level still, cities naturally reflected the regions and republics in which they were situated. Those in the less developed parts of the USSR or with limited industrial development thus had lower living standards than others. Inequalities between Soviet cities were reflected in public perceptions of residential desirability (Sidorov, 1992).

No less significant were inequalities within cities (Fig. 8.3). These have been discussed in Chapter 6. The important point to note is that the source of inequality was basically different from that under capitalism. What mattered in Soviet society was not money, which often enough could not buy goods permanently in short supply, but privilege. Access to housing, special retail facilities, goods in short supply, 'closed' services, were things which (bribery apart) money could rarely buy. Position in the party, the state, a good job in the system of a powerful ministry, were the things which mattered. And it was these things which ministries and organizations used to attract and keep the labour they needed, thus promoting a paternalistic relationship

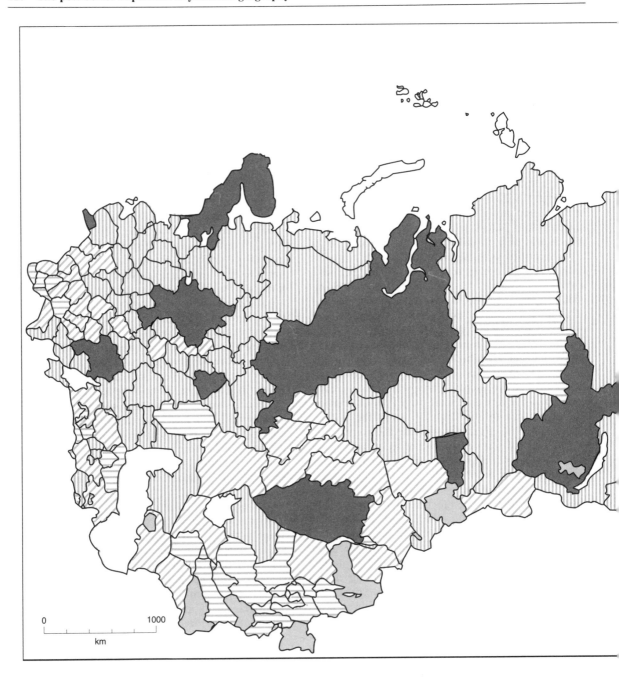

with their employees far removed from that typical of much Western-style capitalism.

Soviet socialism therefore promoted social inequality. But how is one to characterize such a society? Was it merely another type of capitalism with its characteristic forms of social exploitation, or was it a distinctive type of society with its own way of working – a 'distorted'

form of socialism, perhaps? The question is a highly controversial one, but nevertheless ·important since many Western geographers seem to have assumed that Soviet society was capitalist without bothering to look at that society in any detail. The point is, if Soviet society was capitalist, then its cities were capitalist also and lacked any fundamentally distinguishing features.

Percentage of population urbanized

13 to 29

29 to 45

45 to 60

60 to 76

76 to 92

Fig. 8.1
USSR: levels of urbanization, 1989.

This is not the place to go into theoretical debate on this matter, since the issues involved are extremely complex. What we do wish to suggest, however, is that it is unwise to assume, without further investigation, that the current debate in the Western geographical literature about the nature of the capitalist city has automatic relevance to the Soviet city as it was

before 1991. For one thing, for whatever reason Soviet cities do appear to have been very different from those in many other parts of the world (see: French 1979; Shaw 1987). For another, the heritage left to the post-Soviet city is also unusual, as will be suggested by the rest of this chapter. At the same time, Soviet urban society did have some features which are

reminiscent of capitalism. For instance, some scholars have argued that Soviet society was gradually becoming increasingly individualistic. The accent on individual family apartments, the private car, the weekend cottage or dacha, and other features encouraged ever more differentiated styles of life among Soviet citizens. But whether such tendencies reached the stage where individualism became a central characteristic of Soviet urban society, rather than being a feature of more privileged groups, is debatable. For example, unlike the Western city, Soviet cities remained remarkably dependent on public transport.

During the first fifteen years or so after the 1917 revolution, many social thinkers gave their minds to the task of designing the ideal Soviet city. Architects and planners, both Soviet and non-Soviet, joined the

Fig. 8.2
Former USSR: republican capitals and major cities.

quest for a socialist settlement system which would be appropriate to the new type of society now arising in the USSR. That quest was accompanied by much excitement and not a little Utopian optimism. It was a quest which found little sympathy with Stalin. Under his regime, city planning was redirected to the urgent task of supporting the programme of industrialization and away from seeking socialist ideals.

Indeed, planners were brusquely told that, since the USSR was now socialist, its existing cities were already socialist by definition. Planners were therefore ordered to concentrate on practical matters. In response to Stalin's industrialization drive, cities began to grow massively and even chaotically, and city planners faced the task of trying to influence events largely beyond their control. Despite some changes to the

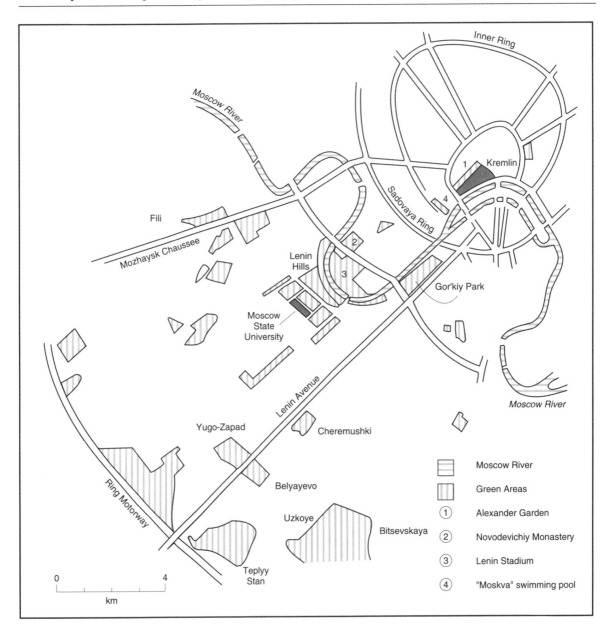

Fig. 8.3
The south-western sector of Moscow: a more sought-after residential area of the city.

system in the post-Stalin period, planners continued to exercise only partial control over urbanization. In consequence, a whole range of problems has been bequeathed to the post-Soviet cities. Some of these have been discussed already: lack of investment capital, lack of co-ordinated planning, shortages of housing, inadequate services, social inequality. Others have only been hinted at: economic stagnation in some cities, alienating and monotonous townscapes, lengthy journeys to work, environmental pollution, inadequate land use control, repressed social and ethnic tensions, poor health, stress, corruption and much else. It is with this Soviet heritage that cities must now face the radically different future unfolding before them.

Post-Soviet urban change

'The train of history is not really a train. The engineer guides it into some station, yet the train arrives somewhere else' (Moshe Lewin 1988).

The Soviet city began to change quite markedly after 1985. At first Gorbachev's reforms were intended to bring about change in a controlled way, within the parameters of the Soviet system. But by the end of the 1980s, it was clear that the processes of change had assumed a life of their own and that the system established by Stalin was doomed. Since then, and more particularly since the break-up of the USSR, cities have been experiencing changes which are so new, radical and dramatic that unfortunately we have as yet no real theory to explain them. The rest of this chapter will focus upon four types of change currently affecting post-Soviet cities: economic change (particularly marketization), political change, the search for national identity, and social change. What is important to emphasize at the outset is that cities in different republics and regions are experiencing different emphases in change. Thus the more Westernized parts of the former USSR, particularly the Baltic states and, less certainly, western parts of the Russian Federation, may find adopting certain European ideas and values less difficult than the Central Asian republics and other southern or eastern regions which may look elsewhere for their models.

The post-Soviet city and economic change

Gorbachev set out to reform the command economy and ended up by destroying it. The resultant economic chaos, falls in living standards and shortages of goods inevitably affected life in the cities where short-time working, queueing and rationing became even more common than they had been before. Inevitably people tried to find ways around the problem: cities by trying to ration and to restrict inward and outward flows of goods; cities, managers and individuals by engaging in barter and semi-legal deals in a tradition which goes way back into Soviet history. The black market flourished and street crime, which had been relatively rare for most of the Soviet period, revived. Even organized crime, which had previously been most notable in some of the southern republics, became much more widespread. Not surprisingly, public discontent mounted and political protest, fuelled by nationalistic and other ideologies, became common. Economic problems were exacerbated by the break-up of the USSR in 1991 which disrupted supplies of raw materials and manufactures travelling between republics.

The breakdown of the command economy has resulted in policies of marketization and privatization, although some republics have pursued such policies with greater enthusiasm than others. For the first time since NEP cities will have privately-owned industry, privately-run shops and services, privately-conducted businesses of all kinds. Already the streets of Moscow and other big cities are becoming lined by international retail outlets, international businesses are opening offices of all kinds, and internationally-owned hotels, restaurants, travel firms and finance houses are making their appearance. It is a vivid example of the impact of the global economy about which geographers now talk so much. But a number of caveats need to be made about this process. In the first place, the process of marketization, which implies not only the privatization of existing enterprises but also the establishment of new, especially privately-owned activities, is likely to benefit the economies of some cities more than others, even in the most rapidly marketizing republics. Capital cities like Moscow, those well-placed close to coasts or frontiers or in strategic locations with respect to international lines of communication, administrative centres for economically well-endowed regions, or cities with other advantages may well prosper more than those in remote locations or with narrow or outmoded economic profiles. Secondly, the impact of marketization will vary in accordance with government policy. Where governments fear the impact of market forces upon towns with only one or two industries (such as the so-called 'company towns' which are often found in the east) or where they wish to preserve militarily-oriented industries, they may well decide to subsidize such activities. Even privatization of such industries may not change this policy.

Privatization is not restricted to the industrial and service sectors. While full land privatization is still restricted in certain republics, various forms of leasing arrangement are likely to lead to a partial marketization of land use. Privatization of property and particularly of state-owned housing (most urban housing being state-owned) is also underway. The latter is being undertaken to encourage citizens to take over more of the burden of providing and maintaining housing from the state and in the belief that a fully-fledged housing market will be more efficient than the state sector in solving housing shortages. Other activities formerly provided by or for the state, including architecture-planning, education, health and welfare, are also undergoing a partial marketization in many cases. Western commentators sometimes talk of the 'commodification' of aspects of urban life in the Western city. An analogous process is now underway in the east.

Western geographers have spoken much of the restructuring of capitalist economies which has been going on since the early 1970s. One version of this talks of a move from a Fordist industrial economy, based upon standardized production and mass markets, to a post-Fordist system of flexible specialization, customized production and segmented markets. Clearly, the former USSR, with its highly protected state-run economy, was long sheltered from the pressures to move in the same direction and remained in an essentially Fordist framework. The price it paid was to lose out in the race for technological supremacy. Now that the post-Soviet republics are being reintegrated into the world economy, the deindustrialization which has been experienced elsewhere is threatening them also. It may take a particularly aggravated form.

Proponents of marketization believe that it will have a salutary effect on all aspects of urban life: both individuals and institutions will benefit by being forced to become more competitive and enterprising. Competition is certainly occurring at every level, and even city governments have now adopted the 'urban boosterism' so common in the West. But clearly there are both winners and losers in this process, and people learn to cope in various ways. Traders, middlemen, dealers and operators in every area of life interlink both legitimate and illegitimate spheres of activity in ways which are reminiscent of the shadow economies of the communist era. Just as the middlemen exploited the gaps in the command economy, so they now exploit the imperfections of nascent capitalism. Corruption, crime and mafia-type activity are rife, taking their cue from the breakdown in law and order which has accompanied economic collapse. Inevitably, working people will try to protect themselves against the negative repercussions of economic change through their trade unions, in strikes and by various forms of political activity. In the last analysis, many try to cope through informal social networks embracing family, friends and acquaintances, much as they did in the communist past. It is through such activities as growing their own food (there has been a huge increase in the number of allotment gardens around Russian cities, for example) that people are creating the safety net which the state has so conspicuously failed to provide.

It is too early to say whether marketization will produce urban landscapes which are similar to those found in Western cities. On the one hand, such processes as the building of city-centre office blocks and retail outlets, and of low-rise, suburban housing estates (already appearing in some places), will certainly enhance the similarities. On the other, much depends upon the economic consequences of marketization. There are many local factors, such as the activity of the state or the role of the informal sector, which may modify the impact of globalization. Thus far the major change has been in the 'feel' of the city, especially in some of the major centres in Russia for example. Whether it concerns the type of cars now visible on the streets, the kinds of goods on display in the shops, or the speed with which people walk to work, it is hard not to notice the differences from life in the Soviet city.

The post-Soviet city and political change

Throughout most of the Soviet period city government was entirely subservient to that of the Communist Party and the state. While some cities, notably Moscow, Leningrad (St Petersburg) and the republican capitals had had some meaningful influence over their destinies, the room for manoeuvre was always limited, and in the case of most towns and cities extremely limited. This meant that the ability of city governments to plan and manage their cities was quite circumscribed, with consequences which have been alluded to already. Under Brezhnev from the 1960s a *modus vivendi* with city and regional governments was worked out: local officials would largely be allowed to enjoy the fruits of office undisturbed, providing they cooperated with the centre in matters it considered important. Thus evolved a cosy, bureaucratic relationship between local party and government officials and the representatives of national authority. It was a relationship which bred complacency, corruption and greed. In the meantime city problems went unresolved.

This unsatisfactory state of affairs, which afflicted every level of Soviet political life, prompted Gorbachev to accompany his economic reforms with political reform. His political changes, which included the democratization of city government, were designed to outmanoeuvre those opposed to his economic policies but in the event they were only partially effective in altering the situation at the local level. With certain notable exceptions, such as some of the big cities like Moscow and St Petersburg, power continued to be exercised locally by elements from the old communist system. This was often because such elements were the only politically and administratively experienced people available and because of the lack of political sophistication among the general populace. The old guard, naturally enough, generally tried to operate as it had always done: aiming to please vested interests. Nevertheless, Gorbachev's policies did have momentous and largely unintended side-effects which wholly altered the overall political context within which cities found themselves. More and more, cities found themselves drifting loose from central control, able on the one hand to determine their own affairs according to their own interests, but on the other hand having to find their own ways in the world without the central support which had been so vital to them in the past.

Since 1991, cities throughout the former USSR have had to come to terms with sweeping change at almost every level and adjust to those changes according to their individual circumstances. In the first place, there are the economic changes which have been discussed above. Responses to these are likely to vary according to economic circumstances in each city: capitals of the newly independent states, for example, might well favour economic policies which will give them the opportunities of participating in the global economy, expanding their service sectors and finding new niches in the world market place; others with economies more obviously reflecting Soviet heavy industrial priorities will probably find a strong constituency favouring propping up parts of the old economic system. Governments of the newly independent states will presumably be forced to take these local circumstances into account in deciding their policies, or run the risk of local disaffection if they do not. A further point is that relations between cities and their respective state governments are complicated by the lack of stable, constitutional divisions of powers in many of them, some state governments being more authoritarian than others, or swinging between greater or lesser degrees of authoritarianism. Disputes about respective rights and powers, sometimes referred to as 'battles of laws', are common. City politics are obviously much affected by the political instability afflicting many of the republics, with strikes, civil and ethnic unrest, and even outright civil war in some places.

Finding a political framework suitable for managing the city's affairs is proving no easier than finding one to manage the state. The problem is that after more than 70 years of communist government, and little or no experience of constitutional government before 1917, it is difficult to agree about either the purpose or the method of government. Because of their particular mode of development since 1917, the post-Soviet republics lack the conventional social divisions found in Western societies, such as that between property-owning and non-property-owning classes. This may be one reason why they have found it so difficult to develop the stable structure of political parties which parliamentary government seems to require. In these circumstances, policies intertwine with personalities and political activity becomes quite unpredictable. A classic case in point has been the struggle for power in Russia between executive and legislative arms of government, both national and also local. In the amorphous political atmosphere surrounding these

struggles, it becomes difficult to say how far they are about policy and how far about naked power, or to interpret them basically as issues of principle.

Some theorists have argued that what is crucial to the development of modernized (in the Western sense of 'modern') societies in the former USSR is the evolution of what is termed 'civil society'. This is a term used to describe voluntary and spontaneous group activities engaged in by individuals without the sponsorship of the state. Needless to say, the opportunities for such 'unofficial' activity were limited under communism, although there is evidence that party control over many spheres of life was weakening towards the end of the period. Some scholars have argued that it is the lack of 'civil society' which inhibits the emergence of Western-type democracy at the present time. Yet others have suggested that long-established Russian traditions of communal democracy, as presently seen in the emergence of neighbourhood committees and similar organizations in some cities, may have the potential to develop 'grass-roots' democratic forms which are unfamiliar in the West.

Certain commentators have argued that one factor which lies behind so many political disputes in Russia and other post-Soviet states is alternative visions of future society. According to this view, the choice is between a more 'liberal', individualistic society, like the Western model, and a more 'conservative', organic society, implying perhaps a 'controlled' type of capitalism. Both models are probably compatible with political authoritarianism, but only the latter with the preservation of significant elements from the command economy. Different republics are likely to take different routes, with city politics evolving accordingly. What seems indisputable is that cities will be managed in a different way from in the past. City governments may possibly be more democratic, but the extent of that democracy is still uncertain. Democracy may be constrained by the political inexperience of the populace and also by traditions of close central supervision of poorly trained or indolent local officials. Cities may also enjoy more autonomy to raise taxes and determine their own affairs. Again, however, the extent of that autonomy remains uncertain, while the financial resources actually available to cities will no doubt vary with circumstances. Above all, cities seem likely to have to learn to operate in a more pluralistic and less predictable environment. Whereas in the past they were obliged to negotiate with the party or the state and its officials, they will now be faced with a variety of often competing individuals, institutions and interests. It is a much more open-ended game, and one for which their communist past has ill-prepared them.

The post-Soviet city and nation building

While the Soviet authorities soon rejected the Utopian social engineering schemes of their early town planners, they gradually appropriated the idea of using the townscape to inculcate social and political values. In essence, this was a traditional approach to planning, with a Russian pedigree stretching back to the reign of Peter the Great. In the Soviet period it was used to educate the populace into the ideas of Marxism–Leninism and to glorify the Soviet state and its leaders. Beginning in the 1920s and 1930s, and continuing with the postwar rebuilding of Soviet cities and beyond, the Soviets manipulated ceremonial spaces, created monumental architecture and composed such signifiers as city, street and building names with the aim of persuading the population to think Soviet thoughts and to behave in the Soviet manner. Under Stalin and later, the city, or at least its central parts, became a sort of theatre for the playing out of ritual dramas and ceremonies like the May Day parades and revolutionary processions. Even where architectural forms were inherited from the past, as in the case of historic Russian churches and ensembles such as the Moscow Kremlin, their meaning was reinterpreted to glorify not religion or the tsars, but the historic achievements of the 'great Russian people'. The tradition of monumental architecture continued after Stalin's death, for example with the renewed fervour for war memorials under Brezhnev. Even when urban architecture assumed modernist and internationalist forms, it was given a socialist realist gloss by Soviet commentators who proclaimed that it reflected the progressive and forward-looking character of Soviet society.

Even under Gorbachev, there was evidence of a retreat from this official ideology. Political slogans began to disappear from buildings, less use was made of propaganda posters, and memorials were removed in some places. There was a certain symbolic irony in the way in which buildings in central Moscow, which had previously carried signs glorifying the Communist Party or 'the great Soviet people', now began to carry advertisements for Pepsi-Cola or McDonald's. The renaming of places became a symbolic battleground, especially in cases such as that of Leningrad (St Petersburg) which had close associations with Lenin himself as well as with the heroic achievements of the 1941–45 Great Patriotic War. The tendency now was for Russian cities to favour their historic, pre-revolutionary names. Non-Russian cities, however, not only de-Sovietized but sometimes de-Russified by inventing new names or by modifying their traditional names so as to divest them of their Russian elements. Such processes have gained further force since 1991 and

the fall of communism. Whether the ultimate result will be to erase from the map and the townscape all the surviving symbols of communism (even tsarist symbols were never entirely obliterated) remains to be seen.

The retreat of communist ideology has naturally left a vacuum which other ideologies are now striving to fill. The most obvious of these is nationalism. While Soviet ideology paid some court to national traditions, its overall effect on the townscape (coupled with other influences such as that of standardization of design) was to reduce the differences between cities located in different republics and regions of the USSR. The last few years have seen the attempt to move away from such uniformity. The assertion of new national identities has had such obvious effects on the townscape as the use of new flags and national symbols, signage in local languages rather than Russian, new memorials, conservation of the past and similar features. The assertion of national identities also affects daily life in the city, as in the sale of books, souvenirs and other artefacts reflecting local traditions, rather than the Leniniana and communist products of the past. A related process has been the revival of religion with the building or reopening of churches, monasteries, mosques and other places of worship, and the use of the various ways of propagating religious belief.

Nationalism and religion may therefore help fill a void for some of those who have lost their communist faith or for others who never had any. But they may do so at a price, for those who do not belong to the national group or share the traditional religion may feel threatened. In extreme cases, the perceived threat may actually be made manifest in ethnic strife and violence. Unfortunately, the post-Soviet republics have experienced all too many instances of this kind already, adding to the pressures for inter-republican migration which particularly began to be felt in the later Gorbachev years and after the Soviet break-up. Inter-republican migration, and in extreme cases 'ethnic cleansing', may thus result in many cities assuming a less varied ethnic and cultural character than they had previously.

Some scholars have argued that the assertion of national or religious identities is a 'premodern' way of adjusting to social change. This is on the grounds that such responses go against the tide of globalization and of the greater pluralism and greater tolerance which, so it is asserted, typify life in the twentieth-century city, particularly in the developed world. The idea that there can ever be a single, overarching 'world view' which can unify society seems, according to this view, more reminiscent of the medieval world or of failed Soviet totalitarianism than of present-day realities. Big, former Soviet cities like Moscow,

St Petersburg and Kiev, have certainly been the stages for the development of a much more pluralistic society than was ever apparent before. Evidence of differing values and mores, of Bohemian styles of life, of crime and violence, of differences of belief and differences in aspiration, are apparent on every hand. Some of this no doubt reflects the excitement of discovery. There is the newness of the possibility of choice in societies in which choice was denied for so long and which have now lost their belief in the old ideology. But the rest reflects the turmoil of a (post)modern world. Evidence of growing pluralism is less apparent in smaller or in more peripheral places, but then that is also true of the West. It does not mean that the periphery will be unaffected. Tolerance, however, is quite another matter. As in the West, many people are fearful of the anarchic implications of such trends, of the decay of 'traditional values'. Many are determined to resist. Just like the West, then, the former Soviet city remains a contested arena in which different values and styles of life coexist and do battle for survival.

The post-Soviet city and social change

Since social change is usually a more gradual process than the economic, political and cultural changes discussed so far, it makes sense to deal with it last. In the final analysis, however, social change is the most fundamental type of change to influence the post-Soviet city, since it will determine the kind of city it eventually turns out to be.

The growing individualism of life in the Soviet city, at least among the most privileged social strata, has been mentioned already. The post-Soviet city seems destined to move even further along this path, particularly in those republics which marketize most quickly and dispense with the remnants of the command economy most decisively. Individual home ownership is becoming a fact of life already, as we have seen. For the majority of city dwellers, this change will be little more than cosmetic since they will continue to live in the same apartment blocks as before. They will, however, have greater freedom to maintain and decorate their properties according to their whim (for which privilege they will undoubtedly have to pay) and to dispose of them to whomsoever they wish, providing they can find alternative accommodation. The fortunate minority may now find it easier to acquire a comfortable country cottage or dacha in addition to their urban apartment, or to replace the latter with a single family house built in the suburbs. As noted earlier, the acquisition of at least a garden plot with associated chalet is now becoming a real possibility for citizens in Russia and other republics. The motor car is also beginning to appear on the shopping lists of more people, with foreign and expensive cars becoming the status symbols of the rich. If the market economy flourishes and increases affluence, there will be ever greater opportunities for people to spend their non-working time, not in the drudgeries of shopping and domestic chores, but in leisure pursuits of their choice. At the same time, if the state continues to withdraw from its welfare commitments (except for providing a 'safety net' for the destitute) and from many types of public provision, so people will be forced to become more dependent on themselves and their families. Whether all this really amounts to individualism in the true sense of the word rather than mass commitment to a new brand of materialism, is left to the reader to decide.

What seems inevitable on current trends is that post-Soviet society will become increasingly unequal. Previous sections have argued that Soviet society was itself unequal in numerous important respects, but what appears to be in prospect now is a new scale of social inequality. Already emerging at the top end of the scale are the new wealthy, those who have made money from recent economic changes. Some of these people are very wealthy indeed, often having acquired their wealth from trade and similar entrepreneurial activity rather than from manufacture. The links between these 'middlemen' and semi-criminal or criminal activities of various kinds are sometimes significant. Beyond these circles are many other groups which have been able to make money through foreign travel or engaging in businesses of different kinds. Much less fortunate are those who for one reason or another have been unable to benefit from the growing market economy: many industrial and service workers (some now facing unemployment) and those on fixed incomes like pensioners and people on welfare. With inflation reaching unprecedented levels recently, all those unable to find additional sources of income other than their traditional employment become very vulnerable. Finally, there are the apparently growing number of unfortunates who have been the victims of the new economy: the homeless, beggars and others, who seem set to form a new underclass.

What is less certain is the longer-term development of this emerging post-Soviet social structure. Will the newly wealthy commercial groups be able to consolidate their position to form a real bourgeoisie? What will be their relationship with the bureaucrats and managers of the old regime? How far will the latter groups, already privileged under the old system, succeed in joining the privileged élite of the new one? (Much will depend on how far elements of the old command–administrative structure are allowed to survive.) Will a 'service class' of professionals and other middle class personnel arise to serve the new system and, if so, how significant will it be? How far will the mass of employees in industry and service activity be able to share in the wealth being created by this new capitalism and how far will they be permitted to retain their dignity and independence? How real will be the 'safety net' of welfare protection for the victims of the system and what will be the opportunities provided for escaping from poverty?

None of these questions can be answered with certainty at the present time; the ultimate answers will probably vary between republics. But it is important to ponder these issues, since the future character of the post-Soviet city will depend crucially upon their resolution. Also important is what happens to 'minorities'. Will married women, for example, continue to bear the burdens of childrearing and domestic duties in addition to their wage earning role, as they so often did in the Soviet city? Will differential unemployment mean women becoming sole breadwinners in some families, with consequent changes in their social status? Will women find new and better niches in the emerging market economy and opportunities which enhance their status and independence? And how far will ethnic and other minorities suffer discrimination and be forced into taking on less desirable employment activities and social roles?

It seems certain that emerging inequalities will be reflected in the social geography of the city and that the latter will also be influenced by Soviet patterns. For example, as suggested in Chapter 6, the housing privatization policies followed by most of the republics have had the effect of 'freezing' existing inequalities by turning housing over to current tenants at little or no cost to them. In other words, the former party boss will continue to enjoy his prestigious city-centre apartment while the struggling teacher or doctor will also be able to 'enjoy' his or her less well-appointed flat in a high-rise block somewhere in the suburbs. Such patterns will no doubt be modified as a result of taxation policies and the operations of an emerging housing market, but it is doubtful if they will change quickly. On the whole it seems likely that currently existing patterns of spatial inequality will be strengthened and deepened rather than radically changed. Thus, city centres may continue to have a residential function, though perhaps a less significant one than in the past. Being located near the shopping and services of the central parts of the city (and such facilities will presumably continue to expand under market conditions), prestige apartments built for occupation by the highly-favoured officials and functionaries of the old regime will still be regarded as desirable properties. Similarly, suburbs or sectors of the city which were the dwelling places of the intelligentsia and white collar workers under the Soviet regime can be expected to continue to retain their superior status, while the opposite will undoubtedly apply to the working class areas situated close to industry. Some parts of the city may become sites for development of single family dwellings for the more affluent. It remains to be seen what will happen to long-distance commuters to the big city, often working in less prestigious occupations: whether they will continue to commute, or whether some will find it possible to migrate to the city (presumably to cheaper areas) once restrictions are lifted.

Will the socially less equal city of the future become as divided and atomized as cities in some other parts of the world? The concept of 'defensible space' – buildings or zones in which special security measures are enforced – certainly applied in the Soviet city to public buildings in general and most particularly to places occupied by leading officials, foreign diplomats and others. But in a post-Soviet urban world, which may be increasingly characterized by crime and violence, it threatens to become even more widespread. Whether the breakdown in law and order can somehow be contained before it undermines current processes of economic and social change remains a crucial problem for the future of the city.

Conclusion

Geographers and others writing about the present-day Western city often argue that it is a city 'in transition'. Some scholars maintain that this transition is affecting the whole of capitalist society and describe it as one from 'organized' to 'disorganized' capitalism (Lash and Urry 1987). The argument is that, as a result of this transition, cities are being subjected to fundamental restructuring, with those at the apex of the urban hierarchy losing manufacturing employment but gaining employment in services (Short 1989; Short et al. 1993). The fortunes of smaller towns and cities vary according to their location and other factors, but everywhere these developments are being accompanied by profound social changes and often by social polarization. At the cultural level, the transition seems to mean a move from belief in the possibility of ordered and rational change, sometimes referred to as 'modernism', to the disordered and uncertain world of 'postmodernism' (Harvey 1992).

Clearly, much of this is also applicable to the transition now underway in post-Soviet cities. But there, and particularly in republics which are marketizing most thoroughly, the transition appears to be both quicker and arguably even more sweeping. No doubt the speed of the change can be partially explained by its lateness. After all, until almost the end of the Soviet era, the Soviet system could be described as an incarnation of modernism, of the belief that a rational and just socialist order could be built by planning from above. The collapse of this belief, at first gradual but gathering pace by the late 1980s, helped usher in the precipitate changes now occurring. But the character of those changes is also influenced by the fact that they are taking place in societies lacking some of the deep-rooted traditions which are characteristic of the West. Marketization is taking effect in societies which were formerly wedded to the command–administrative economic system and were without the traditions or institutions of capitalism which are so well established in the West (Russian capitalism was still in an initial phase in 1917). The result is that the social structure characteristic of capitalism, such as the presence of a bourgeoisie and associated middle class, is hardly apparent in the former USSR. Similarly, the attempt to decentralize political power has been taking place in societies without strong traditions of parliamentary government or a 'civil society' organized enough to contribute to and support national political life. Moshe Lewin writes about the 'short-cut' approach to social change (Lewin 1988). Just as in 1917, when Lenin tried to build socialism in a society which had barely even known capitalism, so now there is the attempt to reconstruct capitalism in a society without some of the apparently important prerequisites.

As the post-Soviet republics open themselves up to the outside world and are restructured by global forces, so their towns and cities will find their own niches in the emerging economic and political order. Such niches will be determined not only by how the individual republics fare but also by how individual regions and cities respond to restructuring; the process of fragmentation which undermined the USSR may yet have a lot further to go. Moreover, the foregoing analysis suggests that restructured cities are unlikely to be identical with those in the West or in other parts of the world; they will undoubtedly bear the hallmarks of their own peculiar histories, in which more than 70 years of communist-type development will loom large. What

will be the end-product of such far-reaching changes? No one can say with certainty. One of the central features of postmodernism is a shattering of previous certainties. It may be that in the post-Soviet cities we are dealing with a special case of postmodernism, a 'postcommunism' with an even higher level of social amorphousness and uncertainty than exists in the West (Kennedy 1992). The only thing we can be certain of is that the character of the post-Soviet city will not be determined by the outworkings of some iron law of history. It will rather be the product of the struggles now going on between social groups and of decisions made by countless individuals.

References

French R A 1979 The individuality of the Soviet city. In French R A, Hamilton F E I (eds) *The Socialist City*. Wiley, New York, pp. 73–104

Harvey D 1992 Social justice, postmodernism and the city. *International Journal of Urban and Regional Research* 16(4): 588–601

Kennedy M D 1992 Social theory after Leninism and communism. *Contemporary Sociology* 21(3): 311–13

Lash S and Urry J 1987 *The End of Organized Capitalism*. Oxford University Press, Oxford

Lewin M 1988 *The Gorbachev Phenomenon: An historical interpretation*. Radius

Ofer G 1976 Industrial structure, urbanization and the growth strategy of socialist countries. *Quarterly Journal of Economics* 90, May: 219–44

Schapiro L 1972 *Totalitarianism*. Macmillan, London

Shaw D J B 1987 Some influences on spatial structure in the state socialist city: the case of the USSR. In Holzner L, Knapp J M (eds) *Soviet Geographical Studies in our Time*. University of Wisconsin, Milwaukee, pp. 201–27

Short J R 1989 Yuppies, yuffies and the new urban order. *Transactions of the Institute of British Geographers* ns 13: 173–88

Short J R, Benton, C M, Luce W B, Walton J 1993 Restructuring the image of an industrial city. *Annals of the Association of American Geographers* 83(2): 207–24

Sidorov D 1992 Variations in perceived level of prestige of residential areas in the former USSR. *Urban Geography* 13(4): 355–73

Zaslavsky V 1982 *The Neo-Stalinist State*. Sharpe, New York

9

Foreign trade and inter-republican relations

Michael J. Bradshaw

For most of the Soviet period, the economy of the Soviet Union was isolated from the global economic system. Stalinist economic strategy stressed the need for maximum self-sufficiency. Then, in the postwar period, the creation of the Council for Mutual Economic Assistance (CMEA), and later the period of East–West *détente*, led to a steady increase in Soviet foreign trade. However, despite being a major industrial nation and a military superpower, the Soviet Union was not a major player in the global trading system. In 1985 the Soviet Union produced 15 per cent of global GNP but contributed only 3 per cent of global trade. Notwithstanding the rapid increase in the value of Soviet foreign trade during the 1970s, by 1988 (the year before the CMEA collapsed) the Soviet Union's share of world exports had fallen to 3.9 per cent and its share of imports to 3.6 per cent. Thus, the Soviet Union's foreign trade was growing at a slower rate than global trade as a whole. But, as the Soviet economy stagnated during the 1980s, imports became an increasingly important component of economic development; in 1988 Soviet imports accounted for 14.6 per cent of national income, or about 12 per cent of GNP (Hewett and Gaddy 1992:16).

As with economic activity in general, control of foreign economic relations remained firmly in the hands of the central authorities. *Perestroyka* heralded a new phase in Soviet foreign economic relations, reforms were introduced that allowed for the creation of joint ventures, and plans were drawn up for the development of Free Economic Zones (FEZs). Reforms introduced to encourage the decentralization of foreign-trade decision making actually promoted greater regionalization and advanced the eventual disintegration of the Soviet Union. Now the post-Soviet republics are having to devise their own strategies for promoting foreign trade and investment.

This chapter begins with an examination of the changing geography of foreign trade in the postwar period and assesses the impact of such trade upon patterns of regional development within the USSR. The following section reviews the policies introduced during *Perestroyka*, including the early development of joint ventures, and considers the consequences of the collapse of the CMEA. As noted above, responsibility for foreign economic relations has now passed to the governments of the post-Soviet republics. The final section examines republican involvement in foreign trade; however, not only are the individual governments responsible for finding their place in the global economic system, but they also have to find ways of managing their trade with one another. Thus the section also examines the character of inter-republican trade, which has now become a crucial component of each republic's foreign trade.

Foreign trade and regional development under central planning

As we have seen in Chapter 4, the Soviet central planning system was relatively successful in developing an industrial economy. The central control of domestic economic activity, and the fact that the rouble was not a convertible currency, meant that a system had to be created to manage interaction between Soviet enterprises and the international economy. The complex bureaucratic system that developed to manage Soviet foreign trade has come to be known as the 'state monopoly of foreign trade'. In market economies it is the individual enterprise that decides whether or not to engage in foreign trade. In a Soviet-type planned economy the central planners determined what should be imported and what should be exported.

The state monopoly of foreign trade

The system that exercised monopoly control over foreign economic relations remained relatively unchanged from the 1930s right up until the reforms introduced by Mikhail Gorbachev in 1987. The intent of the system was to enable foreign trade transactions to be integrated into the economic plans devised by *Gosplan*. Figure 9.1 presents a simplified diagram of the organization of foreign trade in the Soviet Union. The key actors in this system were *Gosplan*, which related foreign trade to the needs of the national economy, and the Ministry of Foreign Trade (which was subordinate to *Gosplan*) which was responsible for implementing foreign trade policy. The State Committee for Foreign Economic Relations was principally responsible for Soviet foreign aid programmes, while the Foreign Trade Bank (*Vneshtorgbank*) handled the financial arrangements for foreign trade. The day-to-day import and export activities were controlled by Foreign Trade Organizations (FTOs), the majority of which were subordinate to the Ministry of Foreign Trade. The system of FTOs was organized by industrial branches. Thus *Avtoexport* was responsible for the export of cars and *Sudoimport* for the importation of ships. The FTOs in Moscow were the point of contact for Western businesses seeking to sell products to Soviet enterprises and it was through the FTOs that Soviet exports were sold on the world market. The FTOs also provided a link between individual enterprises, the industrial ministries and the Ministry of Foreign Trade. In 1985, 90 per cent of Soviet foreign trade was conducted by 50 FTOs which were directly responsible to the Ministry of Foreign Trade, while the remaining 10 per cent of trade operations

were conducted by 9 FTOs responsible to the State Committee for Foreign Economic Relations (Smith 1993: 42).

Not only did the state monopoly of foreign trade enable central planners to use foreign trade to promote policies of the Communist Party, but it also served to isolate the Soviet economy from global financial turbulence. The Ministry of Foreign Trade isolated domestic enterprises from the influence of world prices for imports and exports. Intra-CMEA trade was denominated in 'transferable roubles', an accounting unit devised to balance trade between CMEA member states, while trade with the industrially developed West and the non-socialist developing world was on the basis of convertible/hard currency and barter. An enterprise contributing to Soviet exports would receive payment in roubles on the basis of the domestic price for the good being exported. Distortions within the domestic price system, together with an artificial rouble exchange rate, often meant that the price paid to domestic producers bore little resemblance to the world price that the good was sold for by the FTO. This was particularly true of the export of oil. During the 1970s the domestic oil price did not keep track of the dramatic increases in world price. Thus oil-producing enterprises were only receiving a fraction of the price their oil was being sold for on the world market. Even worse, they saw none of the hard currency generated by that sale since they were paid in roubles. At the same time, enterprises were often overcharged for imported goods, particularly when there was no domestic equivalent of

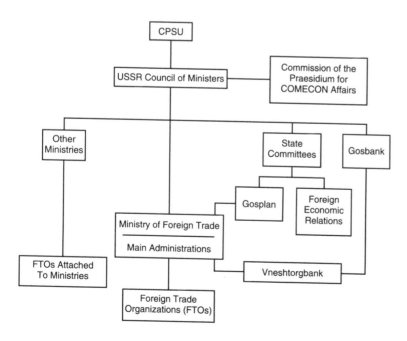

Fig. 9.1
Organization of the Soviet foreign trade system.

the imported item, and therefore no basis upon which to identify a domestic price. Because the enterprise itself was not involved in the process of purchasing imports, it was often the case that inappropriate machinery and equipment were purchased or that they were delivered before the enterprise was ready to install them. There are numerous stories of expensive equipment being left outside in the Siberian winter because it could not be used. The net result of this cumbersome and punitive system was that enterprises had little interest in engaging in foreign trade. Despite these problems, the state monopoly of foreign trade did allow central planners to use foreign trade to address specific national problems. But it is also the case that the isolation of the domestic economy insulated Soviet enterprises from many of the competitive pressures that promoted greater efficiency and technological innovation in open market economies.

Central control over Soviet foreign trade was necessitated by the nature of the Soviet economic system. Now that the Soviet Union and the central planning system are gone, the legacies of the state monopoly of foreign trade present the post-Soviet republics with a number of problems. The first is that under the Soviet system they were never responsible for the foreign trade activities of the enterprises on their territory; trade was always controlled by the FTOs in Moscow. Secondly, because of the nature of the system, enterprises were seldom enthusiastic about developing foreign trade activities. Thirdly, isolation from world prices and competitive pressures now means that the majority of enterprises have little worth exporting and cannot afford imports. Finally, through the centralized control of export earnings, many enterprises, regions and republics benefited from imports that were funded by export earnings generated in other industrial sectors and regions. Now that they cannot call on the central coffers in Moscow, they find themselves isolated from the world economy and unable to finance imports. To appreciate the origins and scale of the problems facing the individual republics, it is necessary to examine in greater detail the nature of Soviet foreign trade prior to the break-up of the Union.

The structure and dynamics of Soviet foreign trade

To understand the dynamics of Soviet foreign trade, it is necessary to examine the commodity structure of that trade. Despite the creation of the CMEA in 1949, it was not until the 1970s that foreign trade started to grow, both in terms of the value of trade and the role that trade played within the domestic economy (Fig. 9.2). The signing of a major agreement

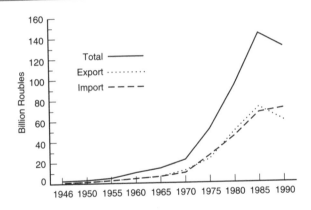

Fig. 9.2
Postwar growth of Soviet foreign trade, 1946–90

between the Italian automobile company Fiat and the Tol'yatti car plant on the Volga is seen as marking a change in official attitudes towards the role of imports in Soviet economic development strategy (see Hanson 1981). However, it was the period of *détente* in the early 1970s and the dramatic increase in world oil prices, brought about by the actions of the OPEC countries, which provided the Soviet Union with the opportunity to finance an expansion of imports.

Despite being a heavily industrialized economy, the trade structure of the Soviet Union was more akin to that of a developing country. The data in Table 9.1 provide information on the commodity structure of Soviet foreign trade. No distinction is made between intra-CMEA trade (conducted on a clearing basis) and trade with the industrially developed West (conducted in hard currency). From these data, it is clear that Soviet exports were dominated by fuels and raw materials. According to Soviet foreign trade statistics, in 1970 fuels and electricity accounted for 15.6 per cent of total exports. By 1975 their share had risen to 31.4 per cent and in 1980 it stood at 33.9 per cent. If one distinguishes between trade with the socialist countries and trade with the industrially developed West the reliance upon energy exports is even more apparent. In 1980 fuels and electricity accounted for 70.1 per cent of exports to the industrially developed West and 39.7 per cent of exports to socialist countries. At the same time, sales of machinery and equipment accounted for 22.8 per cent of Soviet exports to socialist countries and a mere 1.9 per cent of exports to the industrially developed West. The reason for this discrepancy was that Soviet machinery and equipment were of insufficient quality to compete on Western markets, but were exported, through bilateral trade agreements, to eastern Europe and to client states

Table 9.1 Commodity structure of Soviet foreign trade, 1980–89

Commodity (% of total)	1970		1980		1985		1989	
	Export	Import	Export	Import	Export	Import	Export	Import
Machinery, equipment and transport equipment	21.5	35.5	15.8	33.9	13.9	37.1	16.0	38.5
Fuels and electricity	15.6	2.0	46.9	3.0	52.7	5.3	40.2	3.0
Ores, concentrates, metals and metal products	19.6	9.6	8.8	10.8	7.5	8.3	10.5	7.3
Chemicals, fertilizers, rubber	3.5	5.7	3.3	5.3	3.9	5.0	4.1	5.1
Wood and paper	6.5	2.1	4.1	2.0	3.0	1.3	3.5	1.2
Textile, raw material	3.8	4.8	1.9	2.2	1.3	1.7	1.6	1.6
Food and raw materials for food	8.4	15.8	1.9	24.2	1.5	21.1	1.6	16.6
Industrial consumer goods	2.7	18.3	2.5	12.1	2.0	12.6	2.6	14.4
Production services and arms	n.a.	n.a.	14.8	5.9	14.2	7.6	19.9	12.3

Sources: *Vneshnyaya torgovlya SSSR* and IMF *et al.*, 1991 *A Study of the Soviet Economy, Vol. 2.* Paris, OECD: 70–71.

in the developing world. Arms sales were also an important component of this trade. Analyses of OECD trade data show that during the 1980–85 period mineral fuels and energy accounted for an average of 78.4 per cent of Soviet exports to the West. From these data, it seems fair to generalize that during the 1970s and early 1980s Soviet exports earnings were bolstered by windfall gains from the increase in world oil prices (for a more detailed discussion of Soviet oil exports see: Chadwick *et al.* 1987). Further evidence of the rapid, but fragile, growth of Soviet exports is provided by Shmelev and Popov (1990:223) who note that while the value of trade turnover was almost seven times greater in 1985 than in 1970, the physical volume was only little more than twice the size.

Soviet imports were more diversified than exports. Nevertheless, two commodity groups were of greater importance than the others: machinery and equipment, and food. The majority of machinery and equipment imports came from CMEA states. During the 1970s the industrially developed West accounted for about 30 per cent of Soviet machinery imports, but during the 1980s the West's share fell to about 23 per cent. However, Western imports tended to be concentrated in particular sectors. Imports of Western equipment were particularly important to the automotive, chemical and forestry sectors, and imports of large-diameter pipe were essential to the development of the oil and gas industry in Siberia. From the late 1970s, food imports, predominantly grain, from the West increased at the expense of machinery and equipment imports. During the early 1980s machinery and equipment imports accounted for an average of 37.3 per cent of Soviet hard currency imports, while food products accounted for 40.2 per cent of imports.

The commodity structure of Soviet imports suggests three roles for foreign trade: firstly, imports of manufactured goods from eastern Europe bolstered inadequate production of consumer goods and supplemented domestic machinery production; secondly, imports of machinery and equipment from the West, which often embodied technologies unavailable domestically, compensated for problems with the domestic innovation process and helped to develop key sectors and regions of the economy; and thirdly, imports of agricultural products helped to improve the diet of the population and compensated for the inefficiencies of Soviet agriculture. By compensating for the failings of the domestic economy, the expansion of foreign trade put off the need for radical economic reform.

Just as the Soviet Union had benefited from price increases associated with the first and second oil shocks, so it suffered from the third oil shock which brought a rapid reduction in world oil prices. During 1986 the price of oil fell from over US$30 a barrel to around US$10. As a consequence, the Soviet Union saw its oil revenues fall from US$16.1 billion in 1984 to US$7.07 billion in 1986. During 1985 production problems prohibited an increase in the volume of oil exports; however, between 1985 and 1987 the volume of Soviet oil exports to the OECD countries rose by 22 per cent, but the revenue from those exports fell by 38.3 per cent. In 1989 continuing problems in the oil industry resulted in a 3 per cent decline in oil production and an 18.8 per cent decline in oil and oil product exports to the OECD. During 1990 the situation continued to deteriorate, oil production fell a further 5.5 per cent, with exports to the OECD falling a further 15 per cent and exports to eastern Europe decreasing by 27 per cent. In the face of declining

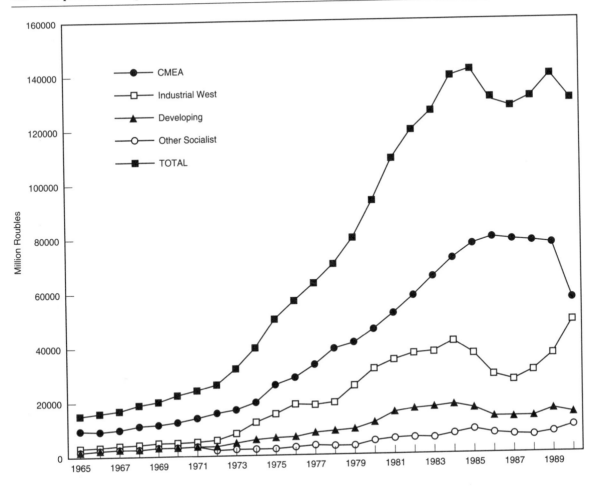

Fig. 9.3
Dynamics of Soviet foreign trade, 1970–90.

hard currency revenues, the Soviet Union switched exports away from the latter region towards hard currency markets (see Fig. 9.3). During 1990 the Soviet Union unilaterally declared that trade with the CMEA would henceforth be conducted in hard currency at world prices. This led to a collapse in intra-CMEA trade. Previously, central control over foreign trade had enabled the Soviet Union to manage its balance of payments. The level of exports was determined by the need to import, if exports earnings fell, then imports would be cut. However, Soviet industry had come to rely on imported machinery and equipment and the population on imported food. Attempts were made to bolster hard currency exports by increasing diamond and non-ferrous metal sales and gold sales were increased. But even these measures could not compensate for the loss of oil revenue. To make

matters worse, during the second half of the 1980s the Soviet Union had increased its level of borrowing. Soviet net debt to Western financial institutions and governments grew from US$10.2 billion in 1984 to US$37.3 billion at the end of 1989 (Smith 1993:158). By 1990 the Soviet Union was facing a serious balance of payments problem. Export earnings had collapsed, gold reserves were depleted and foreign debt and the cost of debt service were growing rapidly. When the Soviet Union collapsed in 1991, these liabilities, by then a debt of US$60–70 billion, were inherited by the post-Soviet republics.

The geography of Soviet foreign trade

Given the dramatic change of fortunes described above, it is not surprising that the geography of Soviet foreign trade has also undergone substantial change. The foreign trade handbooks (*Vneshnyaya torgovlya*) that were published by the Ministry of Foreign Trade (later renamed the Ministry for Foreign Economic Relations) divided the countries of the world into three main groups: socialist (subdivided into CMEA and non-CMEA), the industrially developed capitalist countries and the developing countries. As noted above, this division reflects the different financial systems that used to regulate trade: trade with the CMEA was based on the transferable rouble accounting system; trade with the industrialized West on hard currency; and trade with the developing world on a combination of barter, hard currency and foreign aid, with arms sales playing a major role. Traditionally, about 70 per cent of Soviet foreign trade was conducted on the basis of clearing arrangements, a further 5 per cent on a barter basis, and only 20–25 per cent on convertible currencies. Table 9.2 and Fig. 9.4 use this classification to examine the geographical distribution of foreign trade. In the latter half of the 1980s trade with the West grew in importance at the expense of trade with the CMEA. However, after the collapse of the CMEA trading system, there was a dramatic reorientation in trade. This change was caused not by growth in trade with the West, but by the fact that trade with the West declined more slowly than that with the CMEA. Hence the West's share of Soviet foreign trade increased. Due to the collapse of the Soviet Union, data for 1991 are not available; however, data on Russian foreign trade show the extent of this reorientation (see Fig. 9.5). In 1990 trade with the former CMEA accounted for 42.5 per cent of Russian foreign trade, but by 1992 its share had fallen to 17.3 per cent. Over the same period the OECD's share of Russian foreign trade had increased from 38.3 to 61.3 per cent. This reorientation was due to the fact that between 1990 and 1992 Russia's trade with the former CMEA fell by 80.8 per cent, while that with the OECD fell by 48 per cent.

Classifying trade on the basis of a major trade block hides the fact that most Soviet trade with the 'West' was actually trade with western Europe. In Table 9.3 the foreign trade data have been reworked by major world region. From these data, it is clear that, in terms of the geography of trade, the Soviet Union was oriented towards Europe. Trade with North America was predominantly grain imports, which were financed by a hard currency surplus generated by trade with western Europe. Despite Mikhail Gorbachev's claim that the Soviet Union was part of the Pacific Basin, there is little evidence of any substantial expansion of trade with the Asian-Pacific region. Most of the Soviet Union's trade with the Asian-Pacific region was directed towards client states, such as Vietnam

and North Korea, and had little impact upon the economic development of the Russian Far East. More recently border trade with China has increased rapidly, but trade with Japan remains stagnant. A combined classification based on the major world regions and the commodity composition of trade results in three broad types of trading partner (Fig. 9.6). First, the countries of Europe (east and west) together with Japan, to which the Soviet Union supplied fuel and raw materials in return for manufactures; second, the developing countries to which the Soviet Union exported mainly manufactured goods in return for agricultural products; and third, the United States, Canada, Australia and Argentina, which supplied the Soviet Union with agricultural products, but which imported little in return. Clearly, the fact that most of the Soviet Union's foreign trade was oriented towards Europe should have influenced the impact that foreign trade had upon the regional development within the country. The next section examines the interrelationship between foreign trade and regional development.

Table 9.2 Distribution of foreign trade by major trade bloc, 1970–90 (as a % of total trade)

	1970	1975	1980	1985	1990
Socialist countries					
CMEA*					
Total	55.4	51.8	48.6	55.0	43.8
Exports	54.4	51.1	49.0	55.4	43.2
Imports	57.0	48.3	48.2	54.6	44.4
Non-CMEA					
Total	10.0	4.5	5.1	6.2	6.6
Exports	11.0	4.7	5.2	5.8	6.8
Imports	8.1	4.1	4.8	6.6	6.4
Industrially developed market economies					
Total	21.2	31.3	33.6	26.7	38.0
Exports	18.7	30.0	32.0	25.6	36.0
Imports	24.1	36.4	35.4	27.8	39.7
Developing market economies					
Total	13.4	12.4	12.7	12.1	11.6
Exports	15.9	14.2	13.8	13.2	14.0
Imports	10.8	11.2	11.5	11.0	9.5

Source: *Vneshnyaya torgovlya v SSSR*. various years.

* 1990 data for CMEA do not include East Germany.

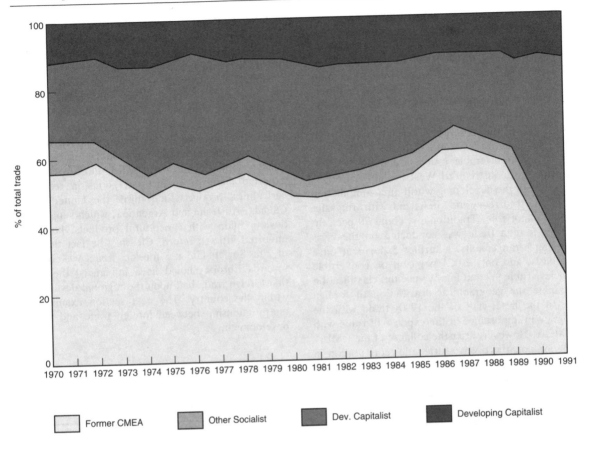

Fig. 9.4

Changing distribution of Soviet foreign trade, by trade bloc.

Foreign trade and regional development

At the outset we must acknowledge that foreign economic relations were of limited economic importance to the Soviet economy. However, because exports were confined to a few commodities and imports were concentrated in particular sectors, it is possible to identify a distinct spatial impact which Soviet foreign trade had upon regional economic development. According to North (1983:99) the 'direct impacts' of foreign trade upon regional development: 'are typified by investments in mines and factories, and in the transport facilities serving them, in order to enable them to export. The allocation of imported factory equipment also constitutes a direct regional impact.' In addition, North identified a number of 'indirect impacts'. For example, the improvement of transport facilities to enable export also changes the comparative attractiveness of the regions served by the transport system. An obvious example of such an indirect impact is the development of the Baykal–Amur

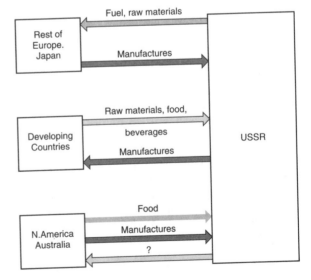

Fig. 9.5

Different types of trading partner of the USSR.

Table 9.3 Geographical distribution of Soviet foreign trade, 1970–90 (as a % of total trade)

Region	1970	1975	1980	1985	1990
West Europe					
Total trade	16.6	23.4	27.0	21.3	32.4
Exports	15.1	22.4	29.7	23.8	32.6
Imports	18.2	24.4	24.0	18.8	32.2
East Europe*					
Total trade	56.9	46.8	46.6	51.5	33.8
Exports	55.3	48.4	46.3	50.6	32.6
Imports	58.7	45.3	47.0	52.5	34.9
North America					
Total trade	1.3	3.9	2.7	2.7	2.6
Exports	0.6	0.6	0.4	0.5	1.0
Imports	2.1	7.1	5.2	4.8	4.0
Asia-Pacific					
Total trade	7.6	7.0	7.2	7.2	9.5
Exports	8.1	5.7	5.3	6.8	10.1
Imports	7.1	8.3	9.2	7.6	8.9
Other					
Total trade	17.6	18.9	16.5	17.4	21.7
Exports	20.9	22.9	18.3	18.4	23.8
Imports	13.9	14.9	14.5	16.4	20.0

Source: *Vneshnyaya torgovlya v SSSR*. various years.

* 1990 data for East Europe do not include East Germany. Soviet trade with East Germany and West Berlin are included in the 1990 figure for West Germany.

Mainline (BAM) railway in the Russian Far East. The combination of domestic and export demand may lead to shifts in patterns of supply or may enable the introduction of new technologies. For example, the prospect of sales to Japan enabled the development of the South Yakutian coal basin, a project that could not have been justified on the basis of domestic demand alone. Finally, North suggests that: 'regions may be affected differentially by changes in the sectoral structure of the Soviet economy, consequent upon the establishment of foreign economic links'. For example, specialization within the framework of CMEA integration led to the development of chemical and petrochemical complexes in the European regions of the Soviet Union, particularly Belarus'.

Because Soviet foreign trade was planned and managed on a sectoral basis via the Ministry of Foreign Trade and the FTOs, information on the regional distribution of foreign trade participation is relatively scarce. A comparison between the commodity structure of Soviet exports and the regional distribution of domestic energy and raw material production suggests that Siberia and the Far East were the most important source of Soviet exports, particularly hard currency exports. A detailed analysis of the impact of East–West trade upon Siberian development (Bradshaw 1992) revealed that the expansion of Soviet exports was made possible by the increased exploitation of Siberia's natural resource wealth. This analysis focused upon five commodity groups: cork and wood, coal, oil and oil products, natural gas, and diamonds and non-ferrous metals, which together accounted for 85 per cent of Soviet exports to the OECD in 1985. According to this analysis, in 1970 exports from Siberia and the Far East to the OECD were worth an estimated US$473 million, which represented 18.6 per cent of total Soviet exports to the OECD. By 1975 their value had risen to US$2.775 billion and their share to 26.8 per cent. In 1985, prior to the collapse of oil prices, the value of exports was estimated to be US$11.251 billion or 53.2 per cent of total exports to the OECD.

The most important exporting region by far was, and still is, Tyumen' Oblast in West Siberia, which in 1985 produced 61.8 per cent of the Soviet Union's oil and 50.2 per cent of its natural gas. In addition, the Far East was a major exporter of forest products to Japan and Yakutia was the only source of natural gem diamonds within the Soviet Union, East Siberia was a major exporter of non-ferrous metals, and both the Far East and East Siberia were important gold-producing regions. The expansion of energy and raw material production to serve both domestic and export markets also resulted in the development of the transport network. Operations on the Northern Sea Route were expanded to supply the West Siberian oil and gas fields, the Noril'sk Metallurgical Combine in East Siberia, and remote mining operations in the Far East. Further south, the BAM and Little-BAM railways opened up the interior of the Far East to possible developments and the expansion of port facilities at Nakhodka was prompted by the growth of Soviet–Japanese trade. Perhaps most importantly, the construction of a network of large-diameter transcontinental gas pipelines enables Siberian gas to be delivered to consumers in the European regions of the Soviet Union and to export markets in Europe. Thus, the direct impact of the expansion of Soviet exports to the OECD was an increased demand for Siberia's natural resources and the development of the transportation infrastructure to enable both resource exploitation and the delivery of energy and raw materials to export markets. At the same time, the growth of CMEA trade served to amplify demand for energy and raw material exports, but the higher share of machinery

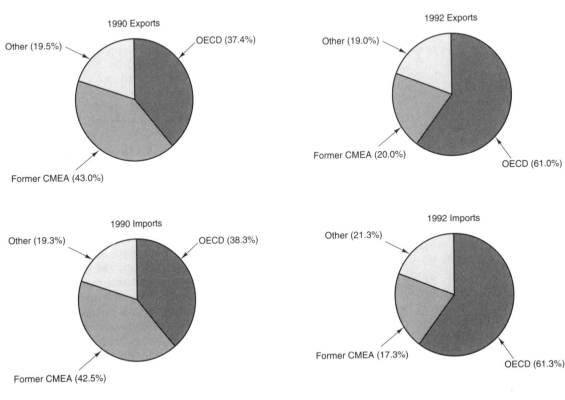

Changing Geography of Russian Trade
1990-1992 (percentage of total trade)

Fig. 9.6
Changing geography of Russian foreign trade, 1990–92.
Source: World Bank (1993:29)

and equipment in Soviet exports to CMEA also meant that foreign trade increased demand for certain types of manufactured goods and chemical products produced by enterprises in the European regions of the Soviet Union.

Analysis of the regional impact of Soviet imports is even more problematic. By reviewing the Western business press, Bradshaw and Shaw (1991) were able to create a computer data base of Western technology sales to the Soviet Union (see also Bradshaw 1991). The pattern of sectoral specialization identified by this analysis was very similar to that revealed by other studies on East–West technology and Soviet economic development (see Holliday 1984). Of the 2500 contracts identified in the period from the early 1970s to the end of 1989, the chemical industry accounted for 17.8 per cent, the automotive industry 14.3 per cent, the oil and gas industry 18.1 per cent and the forestry industry 8.7 per cent. Information on the destination of these imports was available

for 70 per cent of the contracts. The Volga and Central Economic Regions were the most important destinations, followed by West Siberia and the Far East. At the republican scale, the Russian Federation accounted for 64.6 per cent of contracts, followed by Ukraine at 7.4 per cent and Belarus' at 4.0 per cent, leaving 24 per cent for the remaining thirteen republics. Analysis of these contracts also revealed regional specializations. The north-west and the Far East dominated imports to the forestry industry, while imports of chemical equipment were concentrated in the Volga, central, Urals and Belarus' regions. Imports of ice-breaking and ice-strengthened ships from Finland played a major role in the expansion of the Northern Sea Route, Japanese companies provided specialist equipment for the development of the port of Nakhodka in the Far East, and companies in western Europe, North America and Japan provided pipe and equipment for the construction of large-diameter gas pipelines. On the basis of this analysis, Bradshaw and

Shaw (1991:199) concluded that Western technology: 'played a role in the formation of several of the major industrial centres of the European USSR, while imports of pipe and equipment have permitted the majority of the growth to be concentrated in market-oriented locations rather than in remoter resource-producing regions'.

By comparing the geographical impact of exports and imports, it is possible to conclude that the expansion of Soviet foreign trade during the 1970s and early 1980s increased the pace of Siberian resource development, but that the major beneficiaries from imports were industries in the European regions of the USSR. Thus, Siberia's resource wealth was mortgaged to maintain the economic base of the European regions. Most of the inter-regional transfer of foreign trade revenue took place within the Russian Federation. However, republics such as Ukraine and Belarus' were dependent upon Siberia's export earnings to finance their imports. Finally, this transfer of resources and the targeting of imports at specific industries and regions was only made possible by central control over foreign trade. Any measures introduced to reduce the level of central control therefore threatened the ability of importing regions to finance their imports, but promised exporting regions the possibility of benefiting from their export activities.

Perestroyka and the reform of Soviet foreign trade

The reform of the Soviet foreign trade system began in August 1986 when a series of measures was introduced to enhance the role of foreign economic relations in the restructuring of the Soviet economy. These reforms were noteworthy: firstly, because they weakened the Ministry of Foreign Trade's control over foreign-trade decision making, and secondly, because they allowed for the creation of joint ventures between Soviet enterprises and foreign firms on Soviet territory. In this section we examine the decentralization of Soviet foreign-trade decision making, the development of joint ventures, and attempts to create Free Economic Zones.

The decentralization of foreign trade

On 1 January 1987 the Ministry of Foreign Trade lost the monopoly control over foreign trade decision making that it had enjoyed since the 1930s. As a consequence of the legislation introduced the previous August and implemented at the beginning of 1987, 22 ministries and 77 associations and organizations were granted the right to carry out their own export–import activities. In December 1988 the USSR Council of Ministers decreed that as of 1 April 1989, the right to carry out export–import activities would be extended to all enterprises, associations, production cooperatives and other organizations that would be competitive on the international market. By 1990 more than 20 000 organizations had registered. Thus, in a relatively short period of time it appeared that central control over foreign trade had been abolished. However, appearances were deceptive. The central authorities retained control over the bulk of export revenue and controlled the legal environment within which trade was conducted.

Numerous ministerial reshuffles and the reallocation of FTOs to the industrial ministries reassigned control over the right to import, but control over export revenues remained with the Ministry of Foreign Trade. Previously the FTOs of the Ministry of Foreign Trade had controlled 95 per cent of imports from the West; by the end of 1988 its share had fallen to 40 per cent. However, even during 1990 the FTOs of the Ministry of Foreign Economic Relations remained responsible for over 70 per cent of Soviet exports. This was due to the fact that it still controlled the export of energy and raw materials. Thus, while the decentralization of foreign-trade rights enabled enterprises to engage in trade, the central authorities retained control over the revenue necessary to finance trade. This situation was aggravated by a confiscatory tax regime that often left would-be exporters with less than 5–10 per cent of the revenues generated by their hard currency exports. A system of retention quotas was introduced to encourage the export of higher value-added products. Thus if an enterprise exported personal computers it could retain 80 per cent of its export revenue, but if it exported an unrefined raw material it would be lucky to see 5 per cent. Given the realities of Soviet export potential, few enterprises were in a position to benefit from such quotas. In addition, Mikhail Gorbachev introduced a 40 per cent 'Presidential Tax', supposedly to generate income to service the mounting foreign debt. During 1990 and 1991 many enterprises purchased goods from foreign suppliers without access to the hard currency to pay for them. The result was a growing 'non-payments' problem which has made many Western companies wary of doing business in the former Soviet Union.

Given our previous discussion of the geography of foreign trade participation, it is easy to see that this selective decentralization of foreign trade had a distinct regional impact. The industrial enterprises of the European economic core were given the right to engage in trade, while the resource industries of Siberia still saw their export revenue controlled by Moscow. Central control over foreign-trade revenues and the right to determine the legal framework for foreign investment soon became a source of conflict between the Soviet government and the governments of the Union republics. The Union republics sought to gain control over their own economies, and control over foreign trade was seen as a key factor in asserting 'economic sovereignty'. In the so-called 'battle of laws' that ensued the republics created their own foreign investment legislation and asserted that they had legal control over the natural resources of their territories. The conflict was brought to a head by the signing of a five-year agreement, worth US$5 billion, between the Soviet government and the De Beers Diamond Company. An agreement was signed to develop the diamond mines in Yakutia (now the Republic of Sakha-Yakutia), in the Russian Far East. The Russian government protested that the Soviet government did not have the right to enter into an agreement to exploit Russia's natural resources and passed legislation asserting Russia's 'economic sovereignty'. This tactic later backfired on the Yeltsin government as the Republic of Sakha-Yakutia has since declared its own economic sovereignty and has succeeded in negotiating a revenue-sharing agreement with the Russian government in Moscow. Thus, even before the disintegration of the Soviet Union, the Union republics were seeking to gain control over foreign trade activities on their territories.

For the would-be investor the legal chaos that resulted from the 'battle of laws' presented a major problem. Not only was the Soviet government continually changing the legal framework governing foreign trade and investment, but the republics were also introducing their own laws which, they maintained, superseded Soviet law. Foreign companies did not know which legal framework applied to their proposed activities and with whom they should sign agreements. For many companies the solution was to sign agreements with all levels of government and hope for the best. It is against this backdrop of 'chaotic reform' that the Soviet government hoped to entice Western companies to create joint ventures and invest in the future of *Perestroyka*!

The development of joint ventures

In January 1987 two decrees were introduced 'On Comecon joint ventures' and 'On Western joint ventures'. These decrees, and subsequent and numerous amendments to them, form the legal basis for the creation of joint ventures in the Soviet Union and now the post-Soviet republics. The first joint venture was registered in the spring of 1987. By 1 January 1992 the total stood at 5000. However, so far joint ventures have failed to live up to the expectations of both politicians and business people on either side. Prior to examining the development of joint ventures it is necessary to say a few words about the problem of obtaining reliable information on joint-venture activity. First there is the problem of how many joint ventures are registered. In the early phase of joint-venture development, 1987–90, when the Soviet government still exercised control over foreign investment activity, it was necessary for every joint venture to be registered with the Ministry of Finance. However, as foreign trade activity became decentralized and 'republicanized' the registration process became devolved to the republican and then to the local level. Thus, there are no longer any reliable data on the number of joint ventures registered in the post-Soviet republics. The second problem is to determine how many of these joint ventures actually exist. The fact that a joint venture is 'registered' does not mean that it actually exists. For many Western companies the registration of a joint venture represented a 'foot in the door'; for many Soviet enterprises the establishment of a joint venture provided access to tax loopholes. Thus, many joint ventures existed on paper only. To confuse matters futher, the definition of an 'operational' joint venture is not straightforward. The consulting company PlanEcon, who prepared the World Bank study from which some of the data for this section were obtained (World Bank 1992), have defined a joint venture as operational when it has opened a bank account. The State Statistical Committee (*Goskomstat*) declares a joint venture to be operational when it has actually begun to produce goods or to offer services. A third problem is establishing the central activity of the joint venture. Initially, joint-venture registration documents provided fairly succinct descriptions of the main activity of the joint venture; however, when legislation was introduced to limit the export–import activities of joint ventures to their main activities the result was that joint ventures listed as many activities as possible to ensure that they could engage in trade. The net result of these problems is that data on joint-venture activity should be treated with caution. At best they are a rough indicator of the level of foreign investment activity in particular sectors and regions.

During 1987 only 23 joint ventures were registered; by the end of 1988 a further 168 had been registered (see Fig. 9.7). Major revisions to the joint-venture legislation in late 1988 resulted in a substantial increase in activity. Between December 1988 and December 1989, 1076 were registered. By October 1990 the total number of joint ventures stood at 1884 and by January 1992 it had reached 5000 (estimate for CIS only). However, this apparently rapid increase

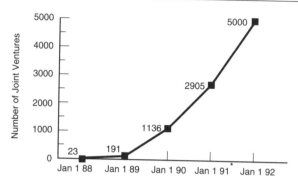

Fig. 9.7
Joint venture registrations, 1988–92.

Following the collapse of the Soviet Union, the fifteen post-Soviet republics have become the focus of investment activity. Table 9.5 provides information on the distribution of joint ventures by republic. Not surprisingly, Russia is the most important location, accounting for almost 84 per cent of total joint-venture registrations in the CIS at the beginning of 1991. The Baltic states are also a favoured location. PlanEcon reports that, following the break-up of the Soviet Union, the number of joint ventures in the Baltic states has increased very rapidly: Estonia had 2200 registered joint ventures in July 1992 compared to 560 registered in July 1991 and 1400 in January 1992 (World Bank 1992:5). According to the UN Economic Commission for Europe (UNECE 1993:15), during the first three months of 1993 the number of foreign investment projects in Estonia grew from 2700 to 4000, and the number of foreign investments in Lithuania and Latvia stood at 2300 and 2800 respectively. Thus, following the collapse of the Soviet Union the Baltic states

in foreign investment is somewhat tempered if one considers the number actually operating. According to *Goskomstat*, on 1 January 1990, of the 1754 joint ventures registered, 703 or 40 per cent were actually operational; by 1 January 1991 the number of operational joint ventures stood at 1027 or 35 per cent. Following the collapse of the Soviet Union, data on joint ventures became even less reliable. *Goskomstat* reported that in October 1991 1570 joint ventures were operational; according to PlanEcon the total number of joint ventures in operation was 3075 (World Bank 1992:3). Whatever the actual number, it is unlikely that more than 40–50 per cent of the joint ventures registered in the post-Soviet republics are operational. Furthermore, most joint ventures are relatively modest undertakings. In late 1990 63 per cent of joint ventures had a capitalization of less than 1 million roubles (about US$100 000 at 1990 exchange rates) and 90 per cent were less than 5 million roubles (US$500 000). According to the joint-venture data base maintained by the UN Economic Commission for Europe in Geneva, at the end of 1990 the estimated total foreign investment represented by joint-venture registrations was US$3.6 billion. Given that only 35 per cent of those ventures were operational, the actual investment could have been as low as US$1.26 billion, PlanEcon estimates it to be US$1.5 billion at the end of 1991 (World Bank 1992:8). Thus, a rather disappointing picture is emerging of very limited foreign investment resulting from the creation of joint ventures. The source of that investment is detailed in Table 9.4. The most important source of foreign investment has been Germany, followed by the United States and Finland. As with East–West trade during the Soviet period, Europe is the most important source of investment. Japanese companies appear to have shown little enthusiasm for creating joint ventures.

Table 9.4 Number of joint ventures by foreign partner

Country	Soviet Union[a]		Russia[b]	
	Number	%	Number	%
Total	2 050		2 747	
United States	247	12.0	398	14.5
Germany	281	13.7	373	13.6
Sweden	66	3.2	212	7.7
Finland	183	8.9	208	7.6
Italy	130	6.3	198	7.2
Austria	115	5.6	164	6.0
Great Britain	112	5.5	122	4.4
Poland	75	3.7	109	4.0
France	70	3.4	90	3.3
Switzerland	78	3.8	79	2.9
Canada	53	2.6	71	2.6
Bulgaria	45	2.2	56	2.0
Hungary	27	1.3	51	1.9
Yugoslavia	32	1.6	50	1.8
China	25	1.2	46	1.7
Japan	33	1.6	43	1.6
India	30	1.5	43	1.6
Australia	30	1.5	32	1.2
Other	418	20.4	402	14.6

Source: World Bank 1992 *Foreign direct investment in the states of the former USSR*. Washington DC, World Bank Studies of Economies in Transformation, No. 5: 20.

[a]   Registered joint ventures as of October 1990
[b]   Operating joint ventures as of 23 April 1992.

have been the most successful of all the post-Soviet republics in attracting foreign investment, although many of these ventures are very small.

Table 9.5 Joint venture activity in the former Soviet Union, 1989–91. Number of joint ventures registered (at the beginning of the year)

	1989	1990	1991	1991 % of CIS JVs.	1991 % CIS foreign trade
Republic					
Russia	141	947	1 971	83.9	72.2
Ukraine	9	83	209	8.9	13.2
Belarus'	3	22	54	2.3	4.2
Uzbekistan	2	12	30	1.3	2.3
Kazakhstan	1	9	16	0.7	2.8
Azerbaijan	2	7	15	0.6	1.3
Moldova	2	13	29	1.2	0.9
Kyrgystan	–	1	2	0.1	0.7
Tajikistan	–	2	3	0.1	0.9
Armenia	3	9	17	0.7	1.0
Turkmenistan	–	1	3	0.1	0.6
CIS TOTAL	163	1 106	2 349	100.0	100.0
Georgia	8	34	77		
Latvia	2	30	162		
Lithuania	4	13	88		
Estonia	14	91	229		

Source: Goskomstat SSSR, 1991, *Narodnoye khozyaystvo SSSR v 1990g.* Moscow, Finansy i statistika: 65.

By 1 July 1992 the number of registered joint ventures in Russia stood at 2649, of which 1353 or 51.1 per cent were actually operational. There was a high level of concentration in Moscow, which was the location of 43 per cent of registered ventures and 39.7 per cent of operational ventures. Other favoured regions in Russia included St Petersburg and the Russian Far East. By the end of 1992 the number of registered joint ventures in Russia was 3252 of which 1250 or 38.4 per cent were located in Moscow. The fall in the dominance of Moscow is probably due to the high levels of crime and corruption in the city and the fact that most manufacturing- and resource-based operations are likely to be based in the provinces. Outside of Russia and the Baltic states, the level of joint-venture activity remains modest. Recently a number of large energy joint ventures have been signed in Kazakhstan and Azerbaijan, but in many republics political instability is reducing investor interest. In the Republic of Georgia, for example, 200 joint ventures are registered, but only four or five are operational.

On the basis of the discussion so far, it should be no surprise that it is very difficult to assess the impact of joint ventures upon the domestic economies of the post-Soviet republics. This is because the functions of the joint ventures are far from clear. Figure 9.8 is based on the PlanEcon data base on joint ventures and gives a general indication of the sectoral distribution of joint-venture activity. From these data, one thing is clear. So far at least, joint ventures have not been a major vehicle for technology transfer to modernize manufacturing activity, as was the case with Western imports during the 1970s. The vast majority of joint ventures are in the

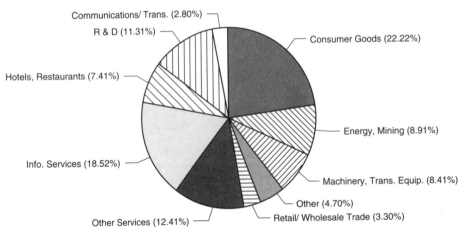

Fig. 9.8
Sectoral distribution of joint ventures, end of 1990, (percentages of total).

service sector. This is one of the reasons why they are concentrated in Moscow. Any visitor to Moscow over the past five years cannot help but have noticed the impact of Western joint ventures on the city's economy. The most obvious case is the arrival of fast-food chains such as McDonald's and Pizza Hut, but a large Western business community is now also supporting the expansion of a financial service sector in the city. At the same time the development of commodity exchanges and the emergence of a business class are providing opportunities for Western retailers to set up shop. Presumably, as economic reform progresses and the legal environment stabilizes, there will be more substantial investment in the manufacturing base of the republics. At present most joint-venture activity outside the major urban areas is focusing upon the development of natural resources, particularly energy resources. Western oil companies are now active throughout Russia, from the Komi Republic in the European North to offshore of Sakhalin Island in the Russian Far East, as well as in Azerbaijan and Kazakhstan. Western mining companies have also shown an interest in mineral deposits in Russia and Central Asia. Uzbekistan's largest joint venture is with an American partner developing the republic's gold industry.

Because of the paucity of reliable information, a thorough analysis of the prospects for joint ventures in the economies of the post-Soviet republics is not possible. Anecdotal evidence from the Western press suggests that the operating environment is still far too risky for many foreign companies to justify large-scale investments. In addition, joint ventures are now only one way in which a Western investor can gain a foothold in the post-Soviet republics. Other options include the creation of wholly-owned foreign subsidiaries, and in some republics foreign investors can now participate in the privatization process, purchasing shares in state enterprises that are undergoing privatization. Most Western companies recognize that the post-Soviet republics represent an enormous market of over 280 million people and that the republics also have substantial raw material wealth. If the conditions were right, then there would be a substantial increase in the level of foreign investment. In such a context, the plan to create Free Economic Zones can be seen as an attempt to create 'safe havens' for foreign investors amid the general chaos.

Free Economic Zones in Russia

Following the rather lacklustre response of Western business to Soviet foreign trade reforms, the Soviet government looked to the creation of Free Economic Zones (FEZs) to attract foreign investors. The success of FEZs in China formed the model for Soviet policy. Special zones would be created and various financial incentives would be provided to attract investment. However, the political environment in the Soviet Union during the 1980s meant that the FEZ initiative quickly became embroiled in the conflict between the central government and regional or local authorities. The Soviet government wished FEZs to operate within the parameters of the partially reformed planned economy. Thus, for example, state enterprises within the zone would still have to act according to state orders. The local authorities wished greater freedom to pursue their own policies. During 1989 numerous local authorities produced plans for the creation of FEZs. In Russia the FEZ became part of the conflict between President Gorbachev and Boris Yeltsin, Yeltsin promising the local authorities greater freedom to develop their FEZs. In the second half of 1990 responsibility for the development of FEZs passed to the republican governments.

Between 14 July 1990 and 27 May 1991 the Russian government granted FEZ status to eleven territories: Vyborg (on the Finnish border), Primorskiy Kray (including Nakhodka), St Petersburg, Kaliningrad Oblast (an enclave between Lithuania and Poland), Chita Oblast, Altay Kray, Kemerovo Oblast (the Kuzbass coalfield), Novgorod Oblast, the Jewish Autonomous Oblast (in the Far East), Sakhalin Oblast and Zelenograd (a suburb of Moscow). These eleven zones together cover an area of 1.2 million square kilometres (7 per cent of the territory of Russia) and have a population of 18.5 million (13 per cent of Russia's population) (Manezhev, 1993:613) There is extreme variation among the zones which stretch from the far western border to the Pacific coast. Each of the zones concentrated on particular economic activities, although all of them wished to develop tourism. Thus, for example, the Kemerovo zone concentrated on heavy industry, while the Zelenograd zone aspired to become a high-technology industrial park. This diversity reflects a lack of a coherent national strategy towards the creation of FEZs. However, once it was perceived that there were benefits to be had from FEZ status, there was a dramatic increase in the number of regions seeking that status.

Following the disintegration of the USSR, the political environment changed once again. In Russia the regions now used the FEZ idea to claim greater powers from the Russian government. By mid-1991, in addition to the 11 FEZs already approved, a further 50 territories were applying for FEZ status. This would have resulted in 61 of the 77 territories being granted FEZ status and would have made a nonsense of the notion that these zones were special territories. The FEZ was fast becoming a pawn in the political power game between Moscow and the regions. Consequently, the Russian government has lost enthusiasm for the FEZ concept. The introduction of radical economic reforms in January 1992, and the ensuing economic crisis, have undermined the ability of local authorities to establish and develop FEZs. Financial incentives are of little consequence in an economy on the verge of hyperinflation. To be successful, international experience suggests that substantial infrastructure investment is required to entice would-be investors. Almost without exception, the local authorities lack the resources to develop the local infrastructure. Thus, at present at least, the FEZ is more a concept than a reality. That is not to say that it is not an important factor in local development. Clearly, regions such as St Petersburg and Nakhodka–Vladivostok are promoting their FEZ status. In future it is likely that the coastal zones of Russia will continue to develop; however, the future of the landlocked zones, such as Kemerovo, must be in doubt. Much depends on the development of new cross-border trading links between the post-Soviet republics and neighbouring states, such as China. At the same time, the other republics are developing their own FEZs to compete with those in Russia. Thus in the future the would-be foreign investor may well be faced with a dazzling array of Free Economic Zones!

External trade in the post-Soviet republics

This final section examines the foreign economic relations of the post-Soviet republics following the disintegration of the Soviet Union during 1991. Of course, in the strictest sense all trade between the republics is now foreign trade. However, in Russia a distinction is now made between the 'Near Abroad', which is the other former Soviet republics, and the 'Far Abroad', which is the rest of the world. Before examining the nature of post-Soviet external trade, it is necessary to consider the legacies handed the newly independent republics by the Soviet system. It is perhaps rather misleading to think of the republics as self-contained economies for, as we have seen, all decisions concerning foreign trade were formerly made in Moscow. Similarly, as explained in Chapter 4, all the major sectors of the republican economies were controlled by the industrial ministries in Moscow. Thus, under the Soviet system the republican pattern of foreign trade activity was merely the geographical expression of decisions made in Moscow. Equally, the pattern of trade between the republics was as much

the result of trade between enterprises in particular industrial ministries, as it was of inter-republican trade. Consequently, when the republics gained their political independence they found themselves in control of 'republican economies' that were not of their making, and responsible for trade relations that reflected centralized decision making, rather than the desires of local industries or republican authorities. On reflection, it is hardly surprising that their immediate response was to restrict inter-republican trade in an attempt to hoard the resources on their territories. A further problem facing the newly independent republics was that the distorted Soviet pricing system had hidden many subsidies. As prices are liberalized it is becoming increasingly apparent who were the contributors and who were the debtors. The following sections examine trends in republican foreign trade activity and interstate (i.e. inter-republican) trade. Any discussion of external trade must be cognizant of the distortions introduced by the domestic pricing system and of the impact of inflation (the monthly inflation rate in Moscow in August 1993 was 26 per cent) upon fiscal measures of interstate trade. Equally, the chaotic situation in the post-Soviet republics means that reliable statistics are hard to come by. Therefore, the following discussion focuses on general patterns rather than on specific details.

The foreign trade relations of the post-Soviet republics

The information in Table 9.6 describes the foreign trade activity of the CIS states in 1991. These data are based on domestic prices; this means that they dramatically understate the value of trade in energy and raw materials. Given what has been said above about the structure of foreign trade, it is clear that such distortions serve to undervalue Russia's contribution to Soviet and CIS foreign trade activity. This is because Russia is responsible for the bulk of CIS exports of energy and raw materials. Despite these distortions, in 1991 Russia still accounted for 72.2 per cent of the total foreign trade turnover of the CIS states. These data also reveal that Russia was the only republic to record a substantial foreign trade surplus. In fact Russia's surplus more than covered the deficits recorded by the other CIS states. This suggests that the Soviet state monopoly of foreign trade masked the fact that Russia's export earnings, or more precisely Siberia's export earnings, financed imports to the other republics. These data make no distinction between hard currency and CMEA trade but, given that hard currency trade was even more biased towards energy and raw materials, it is likely that Russia financed the other republics' hard currency imports. Thus,

Table 9.6 Foreign trade of the CIS in 1991 (million foreign trade roubles according to the commercial rate)

	Exports	Per cent of 1990	Imports	Per cent of 1990	Per cent of total	Balance in 1991
Russia	64 236.5	70.8	44 663.0	54.4	72.2	+19 573.5
Azerbaijan	557.7	62.1	1 410.0	56.4	1.3	−852.3
Armenia	119.5	59.4	1 381.5	80.6	1.0	−1 262.0
Belarus'	2 900.0	58.2	3 418.0	53.1	4.2	−518.0
Kazakhstan	1 354.4	61.6	2 877.0	60.6	2.8	−1 523.0
Kyrgyzstan	79.7	61.3	975.6	57.4	0.7	−895.9
Moldova	270.9	58.7	1 053.9	49.4	0.9	−783.0
Tajikistan	486.0	59.6	797.5	51.6	0.9	−311.5
Turkmenistan	167.3	61.5	697.9	63.2	0.6	−530.6
Uzbekistan	1 131.3	65.5	2 314.3	57.9	2.3	−1 183.0
Ukraine	8 365.2	53.7	11 621.0	61.2	13.2	−3 255.8
CIS total	79 668.1	67.4	71 209.7	61.2	100	+8 458.4
Georgia	355.6	63.0	2 567.5	58.4	NA	−2 201.0

Source: Goskomstat Rossii, 1992, *Kratkiy statisticheskiy byulleten' za 1991*. Moscow: 68.

the break-up of the Soviet Union meant that many of the republics found themselves no longer able to call on Moscow's coffers to finance imports. A combination of declining oil production in Russia and the break-up of the Soviet Union helps to explain the dramatic declines in the value of foreign trade that took place between 1990 and 1991. Recent data from the World Bank suggest that since then the situation has become even worse. The data in Table 9.7 are calculated in US dollars and chart the collapse in trade between 1990 and 1992. Overall, between 1990 and 1992 foreign trade turnover declined by 57.6 per cent. Russia remains by far the most important republic, accounting for 72.4 per cent of total trade turnover, 72.8 per cent of exports and 71.9 per cent of imports. Russia's export earnings declined from US$80.9 billion in 1990 to US$40.0 billion in 1992. However, its trade balance actually improved from a deficit of

Table 9.7 International trade of the post-Soviet republics, 1990–92 (millions of current dollars)

	Exports	Per cent of 1992	Imports	Per cent of 1992	Per cent of total in 1992
Armenia	40	−63.3	95	−88.9	0.1
Azerbaijan	738	2.1	329	−76.7	1.0
Belarus'	1 061	−69.1	751	−85.7	1.7
Estonia	242	22.2	230	−61.1	0.4
Georgia	n. d.	n. d.	n. d.	n. d.	n. d.
Kazakhstan	1 546	−13.0	1 608	−50.5	3.0
Kyrgyzstan	33	−62.9	25	−98.1	0.1
Latvia	429	41.1	423	−74.2	0.8
Lithuania	560	−17.5	340	−78.0	0.8
Moldova	185	−54.3	205	−85.7	0.4
Russia	40 000	−50.6	36 900	−55.5	72.4
Tajikistan	n. d.	n. d.	n. d.	n. d.	n. d.
Turkmenistan	1 083	455.4	545	4.2	1.5
Ukraine	8 100	−39.5	8 900	−44.0	16.0
Uzbekistan	869	−37.5	51 280	−57.6	100.0

Source: calculated from Michalopoulos, C., 1993, *Trade Issues in the New Independent States*. Washington DC, World Bank, Studies of Economies in Transformation, 7: 26.

US$2 billion to a surplus of US$3.1 billion. This was because of swingeing cutbacks in centrally funded imports. Nevertheless, the Russian government has been unable to service the debt inherited from the Soviet Union (Russia has assumed responsibility for the Soviet debt) and continues to ask its creditors to reschedule debt repayments. Despite a generally gloomy picture, there are actually some republics that have managed to increase some aspects of their foreign trade activity. The Baltic states of Estonia and Latvia have substantially increased their exports earnings, while Azerbaijan and Turkmenistan are the only CIS states not to have experienced a substantial decline in exports. The growth of export activity in the Baltic states reflects a reorientation of trade, away from the other former Soviet republics towards Scandinavia and Europe. In the first quarter of 1993 Finland accounted for 23.4 per cent of Estonia's exports, Russia's share was 22.5 per cent and the other CIS states 7.0 per cent. This growth in exports is paralleled by a collapse in trade with the CIS. The situation in Azerbaijan and Turkmenistan reflects the fact that both these republics are energy exporters. Azerbaijan is an oil producer and Turkmenistan exports natural gas. It is only following the collapse of the Soviet Union that these republics have been able to reap the benefits of their energy exports. Overall, the situation is rather gloomy: the disintegration of the Soviet Union seems to have resulted in a collapse of foreign trade and increased economic isolation. One optimistic note is the fact that these figures reflect official trade turnover. A great deal of foreign trade activity is now conducted 'unofficially' to avoid the need to obtain export licences and pay taxes. It may be that foreign trade activity will start to recover once the fiscal and legal environments have stabilized and foreign trade is the business of individual enterprises, rather than governments. The collapse in export activity means that the republics are unable to import machinery and equipment to modernize their economies, and if it were not for foreign assistance they would not be able to import food and consumer goods to satisfy an increasingly frustrated population. This situation has undoubtedly been compounded by the collapse of trade between the republics.

Interstate economic relations

Following the collapse of the Soviet Union, a great deal of attention has been paid to the nature of economic relations between the republics (see Bradshaw 1993; Dienes 1993). Previously it was not an issue because such relations were managed by the central planning system. Unfortunately, that system paid insufficient attention to the economic viability of the various republics. The actions of the industrial ministries in Moscow, together with the Soviet penchant for creating enormous industrial plants, meant that industrial activity was concentrated in a relatively small number of large enterprises. If nothing else, this made the task of central planners much easier. However, it also resulted in a high level of monopoly production, one plant producing all the on–off buttons for TV sets for the entire USSR, for example. This level of concentration, although far from efficient, was manageable while the Soviet Union remained a single economic space. When the Soviet Union disintegrated, the high level of interdependence generated by the central planning system became a major source of political and economic friction. The data in Table 9.8 reflect the situation in 1988, before the economic situation started to deteriorate. The domestic pricing system made it very difficult to calculate the actual value of trade between the republics. As noted above, the system undervalued energy and raw material exports. In addition, price subsidies also distorted the nature of economic exchange between republics. When calculations are made on the basis of world price, Russia and the other energy exporters, Azerbaijan and Turkmenistan, experience a dramatic increase in their terms of trade, while most of the other republics experience increases in their trade deficits. In other words, Russia as the major energy producer was subsidizing the economies of the other republics by supplying them with cheap energy. In such a situation, it is easy to see that Russia's desire to liberalize energy prices threatens to bankrupt the other states. In fact, energy price increases have already resulted in the other republics running up substantial trade deficits with Russia. In response Russia has threatened to turn off the supply of oil and gas unless payment is made.

The gravity of the situation is revealed by the last two columns in Table 9.8, which refer to 1988. One shows the share of inter-republican trade in republican GNP and the other the share of inter-republican trade in total trade turnover. Most of the republics, but particularly the smaller economies, show a high degree of dependence on trade with one another. The republics of the former Soviet Union are actually more integrated than the member states of the European Community. The larger republics, particularly Russia and Ukraine, are less vulnerable, but it was already clear in 1988 that a collapse in inter-republican trade would have a negative impact upon the economies of all the post-Soviet republics. Unfortunately, this is exactly what has happened. The data in Table 9.9 are in constant 1990 prices. From this it is clear that the political disintegration of the Soviet Union has been paralleled by an economic disintegration. The

Table 9.8 Inter-republican trade, 1988 (billion roubles, domestic prices)

Republic	Imports	Exports	Balance	Per cent GNP	Share of total trade
Russia	68.96	69.31	0.23	12.9	58.0
Ukraine	36.43	40.06	3.63	26.9	79.0
Belarus'	14.17	18.22	4.05	44.6	85.8
Uzbekistan	10.62	8.96	−1.66	34.1	85.8
Kazakhstan	13.7	8.3	−5.4	29.5	86.3
Georgia	5.22	5.5	0.28	37.9	86.5
Azerbaijan	4.3	6.3	−2.1	35.4	85.6
Lithuania	6.24	5.43	−0.81	47.3	86.8
Moldova	5.0	4.8	−0.2	45.9	87.8
Latvia	4.6	4.5	−0.1	46.9	86.7
Kyrgyzstan	3.0	2.5	−0.5	39.7	86.9
Tajikistan	3.02	2.0	−1.02	37.7	86.3
Armenia	4.02	3.68	−0.34	47.9	89.1
Turkmenistan	2.5	2.4	−0.1	37.6	89.1
Estonia	3.0	2.7	−0.3	50.1	85.1

Source: *Vestnik Statistiki*, 1990, Ekonomicheskiye vzaimosvyazi respublik v nardnokhozy-aystvennom komplekse, 3: 36 and European Commission, 1991, *European Economy*, 45: 154.

Table 9.9 Inter-state trade of the post-Soviet republics, 1990–92 (millions of constant 1990 roubles)

	Exports 1992	Per cent of 1990	Imports 1992	Per cent of 1992	Per cent of 1992 total
Armenia	1 124	−67.2	1 063	−69.7	1.1
Azerbaijan	2 088	−65.8	1 806	−57.5	2.0
Belarus'	8 526	−50.5	8 200	−44.7	8.7
Estonia	646	−73.8	499	−82.2	0.6
Georgia	572	−90.0	666	−86.5	0.6
Kazakhstan	6 414	−24.0	9 809	−31.5	8.4
Kyrgyzstan	1 086	−55.6	1 407	−55.7	1.3
Latvia	2 220	−55.8	2 186	−53.6	2.3
Lithuania	2 071	−68.5	2 197	−66.2	2.2
Moldova	1 156	−80.2	1 551	−68.9	1.4
Russia	42 464	−43.2	40 148	−40.3	42.8
Tajikistan	1 009	−57.6	1 684	−49.9	1.4
Turkmenistan	2 526	2.3	2 221	−24.0	2.5
Ukraine	18 933	−50.6	23 655	−39.3	22.1
Uzbekistan	2 333	−71.4	2 774	−76.6	2.6
Total	93 168	−50.8	99 866	−47.0	100.0

Source: Calculated from Michalopoulos, C., 1993, *Trade Issues in the New Independent States*. Washington DC, World Bank, Studies of Economies in Transformation, 7: 27

collapse in trade is particularly extreme in republics affected by political instability, such as Georgia and Moldova, and in the Baltic states which are seeking to move out of the post-Soviet arena. The reasons for this collapse are numerous. Some of the most important include the imposition of trade blockades by republican authorities, the collapse of the rouble zone and the introduction of republican currencies, disputes over non-payment, the collapse of industrial production, and the failure of the CIS to create a framework for managing inter-republican trade and payments arrangements. As with foreign trade activity, these statistics reflect the breakdown of the centralized system and the failure of the independent republics to create new institutions in order to manage new economic relations.

Numerous CIS summits have resulted in declarations concerning the creation of an economic union and the rejuvenation of the rouble zone but, as yet, little progress has been made. It may be that only when privatization and marketization have been successfully implemented will new patterns of trade emerge. It is undoubtedly the case that economic reform will be made much easier if the republics can create a framework for encouraging the re-establishment of economic relations between enterprises. But such trade should be based on world prices and comparative advantage, rather than the whims of the politicians and planners in Moscow. For the moment, the most immediate tasks concern the management of energy trade and the creation of a new rouble zone. The CIS now comprises all of the post–Soviet republics except the Baltic states. This analysis of external economic relations suggests that the Baltic states are unlikely to be part of some post-Soviet economic union and that Russia is bound to dominate any union that does emerge. The post-Soviet republics are fast coming to realise that political sovereignty does not automatically bring economic independence.

References

Bradshaw M J 1993 *The economic effects of Soviet dissolution*. Royal Institute of International Affairs, London

Bradshaw M J 1992 *Siberia at a time of change*. The Economist Intelligence Unit, London

Bradshaw M J 1991 Foreign trade and Soviet regional development. In M J Bradshaw (ed.) *The Soviet Union: a new regional geography?* Belhaven Press, London: 165–84

Bradshaw M J, Shaw D J B 1991 East–West technology transfer and Soviet regional development: long term trends and the impact of recent reforms. *Revue Belge de Geographie* **115**: 195–206

Chadwick M, Long D, Nissanke M 1987 *Soviet oil exports: trade adjustments, refining constraints and market behaviour*. Oxford University Press, Oxford

Cole J P 1984 *Geography of the Soviet Union*. Butterworths, London

Dienes L 1993 Economic geographic relations in the post-Soviet republics. *Post-Soviet Geography* **34** (8): 497–529

Gregory P R, Stuart R C 1990 *Soviet economic structure and performance* Fourth edition. Harper Collins, New York

Hanson P 1981 *Trade and technology in Soviet–Western relations*. Macmillan, London

Hewett E A and Gaddy C G 1992 *Open for business: Russia's return to the global economy*. The Brookings Institution, Washington DC

Holliday G D 1984 *East–West technology transfer: a survey of sectoral case studies*. OECD, Paris

Manezhev S 1993 Free Economic Zones in the context of economic changes in Russia. *Europe–Asia Studies* **45** (4): 609–25

North R N 1983 The impact of recent trends in Soviet foreign trade on regional economic development in the USSR. In Jensen R G, Shabad T, Wrights A W (eds) *Soviet natural resources in the world economy*. Chicago, University of Chicago Press: 97–123

Shmelev N, Popov V 1990 *The turning point: revitalizing the Soviet economy*. I B Tauris & Co. Ltd, London

Smith A 1993 *Russia and the world economy: problems of integration*. Routledge, London

UNECE 1992 *East–West Investment News*. No. 2, Summer

World Bank 1992 *Foreign direct investment in the states of the Former USSR* The World Bank, Washington DC

10

Fifteen successor states: Fifteen and more futures?

Denis J. B. Shaw

There can be little doubt that future historians will regard the break-up of the Soviet Union towards the end of 1991 as one of the great geopolitical transitions of the twentieth century. What had begun as an attempt to reform the communist system and restore to it some of the economic dynamism it had known in the early days ended in political turmoil. Not only did the USSR's communist empire in eastern Europe collapse, but Moscow's own dominions disintegrated. Republics which in some cases had been created at the behest of the Kremlin became fully independent states. Borders which had originally been drawn to suit the convenience of Moscow suddenly assumed international significance. It is still too early to assess the long-term consequences of these events. What can be said is that the successor states were ill prepared for the suddenness of their change in status and for the magnitude of the problems it has brought in its wake.

This final chapter will examine the major issues now facing each of the successor states individually, since each state faces problems which are to some degree unique. The development potential of each republic will also be commented on, with particular reference to resources. Before addressing these questions, however, it will be helpful to conclude the arguments of the previous chapters by summarizing the major problems which have been bequeathed to the fifteen republics by over 70 years of Soviet-type development.

Potentially perhaps the most explosive issue facing the fifteen states is the ethnic one. The Soviet Union attempted to solve the ethnic problem by establishing republics and autonomous areas for important nationalities, but the careless way in which the boundaries were drawn up plus the ethnic mixing which accompanied industrialization mean that all the states now contain sizeable ethnic minorities. There is considerable potential for conflict within each republic as well as between republics. The problem for each government is to secure the integrity and independence

of its republic without antagonizing its minorities and its neighbours. Unfortunately in some cases the latter has proved impossible to achieve.

Another grave set of problems are the economic ones. Every republic has inherited an economy which was developed to meet Soviet requirements in heavy industrialization and militarization. Industries developed for these purposes are generally ill suited to the competitive environment of the global economy. Moreover, they are usually extremely resource-intensive and polluting, with unfortunate consequences for human health and the environment in general. Human health and the environment have suffered as a result of the ruthless way in which the Soviets emphasized economic growth and military strength. A further point is that republican economies were developed in the closest possible interdependence with other parts of the USSR and there were many imbalances in and inequalities between different regions and republics. Newly erected barriers between republics are exacerbating what would have been severe problems of restructuring in any case. Economic restructuring means deindustrialization, growing unemployment and the need to rebuild many parts of the infrastructure. Although all the states are proceeding to marketize, it remains to be seen how thoroughly and with what success they will pursue this goal.

Soviet-type development produced many types of spatial inequality, though these are often rather different from those typical of capitalism. One of the most pressing issues is the inequality between town and country. All the republics have inherited an agricultural economy in sore need of restructuring and investment, and an acute problem of rural underdevelopment. In the case of the Central Asian states there is the added problem of a rapidly growing population.

Finally, Soviet-type development left the republics ill prepared for the responsibilities of statehood. Development under the tutelage of Moscow meant that, outside Russia, there was no political class with experience of the highest levels of government, and little understanding in any republic of democratic forms or of market conditions. Republics must now learn to make their own way in the world with little enough guidance as to how to do so successfully.

In addition to all this, each successor state faces unique challenges. It is this issue which must now be addressed.

The fifteen successor states: challenges and opportunities of development

The fifteen successor states vary considerably in their size, resource potential and culture. It would be possible in a discussion of the states and their respective futures to deal with them in almost any order. However, there is a certain amount of logic in treating them geographically by group and this is what will be done here, leaving the Russian Federation as

the biggest republic until last. Some key indicators of the varying characters of the republics will be found in Table 10.1

As noted in Chapter 1, the *western republics* of the former USSR came originally into the Russian orbit mainly in the seventeenth and eighteenth centuries, and they were all previously exposed to varying degrees of European influence. They are located in territories with comparatively moderate natural environments and their rural regions are among the most densely settled in the former USSR, especially in the south-west. However, industrialization has exposed them to the problems of resource depletion, since they lack the abundant minerals and fuels of the former USSR's eastern territories. Given the geographical location of these republics, their territories have frequently been fought over as they have all too often become a battleground between east and west.

Among the western republics the *Baltic states* of Estonia, Latvia and Lithuania, form a distinct group

Table 10.1 The fifteen post-Soviet republics: key social and development indicators

	1	2	3	4	5	6	7	8	9	10	11
Russia	17 075.4	76.2	51.2	81.5	74	118.6	60.7	45.8	90.9	77.3	55.4
Ukraine	603.7	2.7	17.9	72.7	68	89.7	18.1	22.6	0.9	3.9	24.3
Belarus'	207.6	0.9	3.5	77.9	67	116.4	4.1	5.8	0.3	0.0	0.0
Uzbekistan	447.4	2.0	7.1	71.4	40	47.9	2.8	4.9	0.4	5.2	0.8
Kazakhstan	2 717.3	12.1	5.8	39.7	58	73.9	3.7	6.6	4.2	0.8	18.7
Georgia	69.7	0.3	1.9	70.1	56	85.8	1.5	1.5	0.0	0.0	0.2
Azerbaijan	86.6	0.4	2.5	82.7	54	70.9	1.7	2.0	2.2	1.4	0.0
Lithuania	65.2	0.3	1.3	79.6	69	108.6	1.4	2.2	0.0	0.0	0.0
Moldova	33.7	0.2	1.5	64.5	48	118.6	1.2	2.2	0.0	0.0	0.0
Latvia	64.5	0.3	0.9	52.0	71	118.6	1.2	1.4	0.0	0.0	0.0
Kyrgyzstan	198.5	0.9	1.5	52.4	38	53.4	0.7	1.2	0.0	0.0	0.5
Tajikistan	143.1	0.6	1.8	62.3	31	45.4	0.6	1.2	0.0	0.0	0.1
Armenia	29.8	0.1	1.2	93.3	68	78.0	1.0	0.7	0.0	0.0	0.0
Turkmenistan	488.1	2.2	1.3	72.0	45	61.4	0.5	1.1	1.0	11.3	0.0
Estonia	45.1	0.2	0.5	61.5	72	118.0	0.7	0.8	0.0	0.0	0.0

Sources: Narodnoye khozyaystvo SSR various years, Finansy i statistika, Moscow; *Natsional'nyy sostav naseleniya SSSR* 1990, Finansy i statistika, Moscow; Sagers, M., 1991, Regional aspects of the Soviet economy *PlanEcon Report* **7**(1–2), 15 January.

Key
1   Area (thousands of square kilometres)
2   Percentage share of total USSR territory
3   Percentage share of total USSR population, 1991
4   Titular nationality as percentage of total population, 1989
5   Percentage of population urbanized, 1991
6   NMP per capita, 1988 (USSR=100)
7   Percentage share of USSR industrial output, 1985
8   Percentage share of USSR agricultural output, 1985
9   Percentage share of USSR oil and gas condensate production, 1989
10   Percentage share of USSR natural gas production, 1989
11   Percentage share of USSR coal production, 1989.

(see Fig. 10.1). These regions were acquired by Russia in the eighteenth century, before which they had been exposed to Germanic, Scandinavian and Polish influences in particular. All three republics were independent states from soon after the 1917 revolution until 1940, when they were forcibly annexed by Stalin. This is probably why they were pioneers in the separatist movements of the Gorbachev years. Despite their similarities, the three republics have distinctive characters, with the Estonians speaking a language close to Finnish and the other two peoples speaking languages belonging to the Baltic group. In religion, the Estonians and Latvians are traditionally Protestant, the Lithuanians Catholic. In the Soviet period the three enjoyed higher living standards than most of the other Soviet republics.

Estonia

The northernmost of the three Baltic states experienced considerable Russian immigration after the Second World War, as noted in Chapter 3. This was partly because of Soviet restructuring of the Estonian economy to suit Soviet needs. Restructuring meant considerable emphasis being placed on such sectors as fuels and machine building. It was difficult to recruit the labour needed for these expanding industries locally given the low Estonian birth rates (and given the forced deportation of numerous Estonians to other parts of the USSR), and large numbers of Russians were thus moved in. Considerable numbers congregated in the north-eastern part of the republic, particularly around Narva close to the Russian border. Russians were also attracted to Estonia by the relatively high living standards. Altogether, at the time of the 1989 census, Russians constituted almost one-third of the Estonian population, provoking fears for the integrity of Estonian culture. This is the main reason why Estonia opted for a restrictive definition of citizenship, as noted in Chapter 3, and this together with arguments over the withdrawal of Russian troops from Estonian soil has not helped relations with the Russian Federation. There is also potential for border disputes: there have been calls for the Narva region to be granted autonomous status, while the Estonians have made claims on territory now in Russia which belonged to prewar Estonia.

As noted in Chapter 9, Estonia has been relatively successful in reorienting its trade away from the former USSR towards the West and can rely upon traditional economic strengths such as light industries, food processing, agriculture and certain types of engineering. Unlike the other Baltic states, it has considerable fuel resources of its own in oil shales

and can thus to some degree avoid costly imports. Estonia was the first post-Soviet republic to introduce its own currency in June 1992, and has resolutely refused to join the CIS. It evidently wishes to see its future as associated with Scandinavia and central and western Europe. However, despite rapid marketization, Estonia continues to suffer acute problems of economic restructuring.

Latvia

Latvia has many of the same problems as Estonia and a rather similar economic profile. Like Estonia, it has a large Russian minority (34 per cent in 1989) and there were also around 120 000 Belorussians and 90 000 Ukrainians at the last census. In fact, Latvians formed only 52 per cent of the total population in 1989 and its citizenship law is even more restrictive than that of Estonia. There have been disputes over the presence of Russian troops on Latvian territory (a relict of the Soviet era) and over the disposition of the Russian border.

Latvia does not have Estonia's fuel resources and has traditionally been a big importer of energy, especially from Russia. Currently there are attempts to diversify supplies. In terms of its contribution to NMP, Latvia was more dependent on industry than Estonia in 1991, and there was a greater reliance on engineering and metalworking industries (with a defence-related sector). However, Latvia also has a long tradition of light and food processing industries. Like Estonia, Latvia is westward-looking and hopes to benefit from its traditions of skilled engineering and craftsmanship (which were also valued by the Soviets) in restructuring its economy.

Lithuania

Lithuania is to some degree the most distinctive of the three Baltic states and has closer historic associations with Poland, Ukraine and Russia. It is mainly Catholic and has therefore had higher birth rates than the other republics. Until recent times it was also industrially less developed. When rapid industrialization did take place after about 1960 it occurred as a result of Soviet determination to make better use of the local labour supply and there was less need to recruit Russians. Thus Russians constituted only about 9 per cent of the population in 1989, and the Lithuanians about 80 per cent. There is also an important Polish minority (about 7 per cent) and the Poles have an historic claim to the Lithuanian capital, Vilnius. For their part Lithuanians have advanced claims to Kaliningrad Oblast, the Russian exclave which was

Fig. 10.1
The three Baltic states and Belarus'.

formerly the northern part of German East Prussia, annexed by Stalin in 1945. The claim is based upon historic association rather than ethnicity, since the territory is now populated mainly by Russians. The Lithuanians are, however, concerned about Russian transit rights to the territory and about the presence of the Russian military there.

Like the other Baltic states, Lithuania has pursued a westward-oriented course, declaring its outright independence of the USSR as early as March 1990 and flatly refusing to join the CIS. It left the rouble zone in autumn 1992. However, the government elected in October 1992 has been slightly less intransigent on relations with the east and Lithuania has been somewhat slower to marketize than its northern neighbours. It lacks energy resources (despite some domestic oil) and was traditionally very dependent on oil and gas supplies from Russia and on nuclear power. Problems with the latter after Chernobyl' and the rising price of Russian energy badly affected the Lithuanian economy, though the country seems now to be changing its mind about the undesirability of nuclear energy (Marples 1993). In profile, the economy is similar to those of Latvia and Estonia with a larger dependence on agriculture. The country tends to see its future as lying in central and western Europe but, like the prewar republic, it may find it necessary to maintain economic good relations with the east.

The remaining three western republics are: Belarus', Ukraine and Moldova.

Belarus'

Belarus', like the Russian Federation and Ukraine a Slavic country, was the fifth most populous republic of the former USSR but the third (after Russia and Ukraine) for industrial production and fourth (after the same two plus Kazakhstan) for agricultural output (1985 NMP data: Sagers 1991). Belorussians constituted 78 per cent of the population in 1989 and the Russians were the second largest group at 13 per cent. The Russian percentage had grown considerably in recent decades (it stood at just over 8 per cent in 1959) as a result of the rapid industrialization of the period. Numerically the third most important group was the Poles (at 4 per cent) who are long-standing residents particularly concentrated in Grodno Oblast in the north-western part of the republic. Belarus' has experienced less ethnic tension than elsewhere and this may continue as long as it maintains good relations with its neighbours (see Fig. 10.1).

Culturally, Belarus' is in a transitional position between West and East, the western part of the republic tending to look westwards for historical reasons (for example, most of it was part of Poland before the Second World War) and the eastern part eastwards. This tension is to some degree reflected in the country's foreign policy outlook, with parts of its leadership looking towards Russia in aspiration and other parts towards central and western Europe. Belarus' was one of only four post-Soviet republics inheriting a nuclear capability, but has been somewhat less concerned than Ukraine about a possible threat to its security coming from the East. During the Soviet period, Belarus' experienced a relatively high degree of linguistic and cultural Russification.

The republic is not particularly well endowed with natural resources, including energy resources, and has been very reliant for its raw materials and energy on Russia and its other post-Soviet neighbours. Belarus' suffered particularly from the consequences of the Chernobyl' disaster which contaminated as much as one-fifth of its land including some of the best arable and livestock regions. Despite this, the energy shortage has stimulated a renewed interest in nuclear power. The country's heavy industrial legacy means that there is a powerful lobby in favour of a cautious approach towards marketization. Caution is also supported by the well-organized labour movement. The country has thus taken a fairly conservative stance on reform.

Ukraine

Ukraine was the second most populous republic of the USSR and easily the second most significant producer of industrial and agricultural output after Russia itself. Spatially, it was the third biggest republic after Russia and Kazakhstan but occupied only 2.7 per cent of Soviet territory. Even so, it is very large on the European scale, being Europe's biggest country after Russia (see Fig. 10.2).

Ethnically, Ukraine is quite diverse. Ukrainians formed 72.7 per cent of the population in 1989 with Russians the largest minority at 22.1 per cent. With the single exception of Crimea, Ukrainians were a majority of the population in every region but unfortunately for the country's cohesiveness the Russians are particularly well represented in the east and the south (Fig. 10.3). As noted in Chapter 1, this is partly the consequence of the way the steppe was settled originally, and partly of later periods of industrialization (including the Soviet period). Chapter 3 made the point that the large numbers of Russians in industrial regions of the east, including the Donbass coalfield, have encouraged considerable Russian speaking even among Ukrainians in the area (see also Shaw and Bradshaw 1992). This underlines the region's cultural distinctiveness within

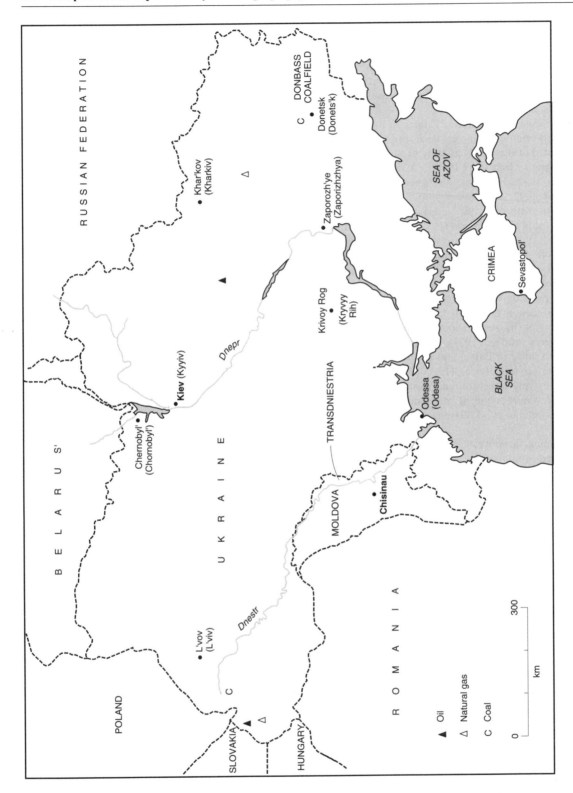

Fig. 10.2
Ukraine and Moldova.

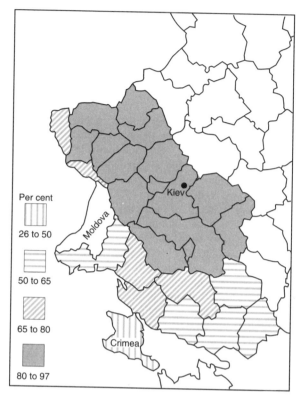

Fig. 10.3
Ukraine: distribution of ethnic Ukrainians, 1989.

Legend:

Per cent

26 to 50

50 to 65

65 to 80

80 to 97

Ukraine. Fears about the possibility of future forced Ukrainianization coupled with concerns about the impact of marketization on the coal mining and heavy industries of the Donbass and adjacent regions have stimulated calls for autonomous status (see Ch. 3).

Because of Ukraine's complex history, there is considerable cultural variation across the territory. Thus the east has long historical associations with Russia, the central part rather fewer, while parts of the west only joined the Soviet Union around the Second World'War. The west is the home of Ukrainian nationalism which has a history far longer than that of Belarus'. Parts of the west are also traditionally given to the Uniat (Ukrainian Catholic) faith which is different from the Orthodoxy of the centre and the east. Some areas in the west are home to minority ethnic groups. Thus Moldovans and Romanians, who are ethnically if not culturally identical, live in considerable numbers in Chernivtsi Oblast in territory which belonged to prewar Romania. Some groups in Romania itself have advanced claims to this territory and to southern Bessarabia, part of Odessa Oblast, and this could complicate future relations between Ukraine on the one hand and Romania and Moldova on the other. In the far west, in Transcarpathia, is a small Hungarian minority and a far larger Ruthenian population which claims to be ethnically distinct from the Ukrainians. Some Ruthenians are separatists who may not be satisfied with Ukraine's generally hospitable policy towards minorities (Stewart 1993).

Crimea is a special case. As noted already, it is the only region of Ukraine where the Ukrainians are a minority and the Russians a majority. This fact was recognized by its being granted (autonomous) republican status in February 1991. However, this special status is insufficient to satisfy some Russian nationalists in Crimea and also in Russia itself who resent Crimea's transfer to Ukraine in 1954. The latter act was accomplished by Khrushchev, supposedly in commemoration of the 300th anniversary of Ukraine's union with (or as Ukrainian nationalists term it, annexation by) Russia. The location of the naval base of Sevastopol', home to the Black Sea fleet, on the peninsula, greatly complicates the issue, since both Ukraine and Russia have ambitions as Black Sea powers. A further problem is the Crimean Tatars who were expelled from the peninsula to Central Asia by Stalin but who have recently been returning to their former homeland. There are many difficulties over their political status and the restitution of their property.

Being a large republic, Ukraine has many resources including fertile agricultural land, coal, iron ore, natural gas and manganese. However, resources were depleted by its rapid industrialization in the Soviet period and Ukraine became more and more dependent on the energy and raw materials of the east. Its industries also suffered from the diversion of investment funds into the development of eastern resources. Its current industrial profile reflects that of the former USSR as a whole with a pronounced accent on mining, ferrous metallurgy, chemicals, heavy engineering and defence-related activities. Much of this industry requires radical restructuring and modernization. However, with a large vested interest in favour of preserving traditional economic activities, Ukraine has been conservative on economic reform, as noted in Chapter 4. Industry and agriculture have suffered severely, not least because of the uncertainty and rising costs of energy and raw material supplies from Russia and other republics.

Ukraine has gained rather than lost territory in the twentieth century and this has made it eager to maintain the territorial *status quo*. It has been particularly fearful of any Russian ambitions for frontier revision and this is one reason why it was at first reluctant to abandon its nuclear weapons. It has also been a somewhat unwilling member of the CIS. However, as noted in Chapter 9, it has been unable to escape interdependency with Russia. Internationally, it has cultivated its relations with central Europe and Turkey, but those with the West have often been made difficult by its cautious approach to economic reform and its stance on nuclear weapons.

Moldova

The small republic of Moldova is situated mainly in Bessarabia, territory which belonged to Romania in the interwar period (Fig. 10.2). It has a particularly difficult ethnic situation since Moldovans constituted only 64.5 per cent of its population in 1989. There are significant minorities of Ukrainians (13.8 per cent) and Russians (13 per cent). The problems of the Transdniester region were discussed in Chapter 3. There have also been difficulties with the Gagauz and Bulgarian minorities, mainly in the south. Ethnic relations have been complicated by the desire of some Moldovan nationalists, and of some nationalists in Romania, for reunification with that country.

Moldova is noted for its rather dense rural population and relatively low urbanization. Its economy has been oriented towards agriculture, food processing, light industries including textiles, and consumer goods production rather than heavy industry. Because of its great dependence on other parts of the former USSR for energy and raw materials and to provide markets for its food production, it has suffered considerably from the disruptions resulting from the Transdniester problem and from the Soviet break-up. Moldova benefited from hidden Soviet subsidies on its energy imports. It has been attempting to cultivate new suppliers and markets in Romania, Iran and elsewhere.

Because of its geographical location in the south-west and its agricultural orientation. Moldova is sometimes counted among the former USSR's southern republics, with generally lower development levels than those to the north (see Ch. 6). The other southern republics are: the Transcaucasian ones, Kazakhstan, and the four republics of Central Asia.

The *Transcaucasian republics* of Georgia, Armenia and Azerbaijan, together with neighbouring parts of Russia's North Caucasus economic region, are situated in the most ethnically and culturally diverse territory of the former USSR (Fig. 10.4). Unfortunately, this has made it one of the most difficult regions of ethnic conflict in the post-Soviet arena.

Georgia

Georgia has a very diverse population, the Georgians forming 70 per cent of the total in 1989. Russians have been returning to their homeland for several decades. There are small minorities of Armenians (8.1 per cent) and Azeris (5.7 per cent), and the fact that many live near to border regions provides some potential for inter-republican conflict. Politically more significant has been the Abkhaz minority in the north-west, which in 1989 formed only 18 per cent of its autonomous republic's population. The Abkhazy have been fighting for independence and there has been a large out-migration of Georgians from the area. Unfortunately for Georgia's economy, Abkhazia lies astride its main communication route into Russia. There has also been sharp ethnic conflict over the former autonomous oblast of South Osetia. The South Osetians have been struggling for union with the North Osetians in the Russian Federation, but their autonomy was abolished by the Georgian parliament in December 1990 in retaliation for their unilateral declaration of independence. In 1989 the South Osetians formed two-thirds of their autonomous oblast's population.

Quite apart from the ethnic problem, Georgia is also divided regionally. President Gamsakhurdia, who was elected to office in May 1991 but then forced from power in January 1992, subsequently found very strong backing in his native region in the west.

The Georgian economy has been noted for its agriculture, food processing and light industries, as well as for production of some minerals such as manganese. A thriving second economy long provided considerable wealth, but this is now dissipating in the wake of post-Soviet troubles. The republic has suffered severely from shortages of energy, raw materials, water, food and other necessities. The loss of the Soviet market for its specialized agricultural production has also been a severe blow. In the future Georgia may well be able to make more use of the resources locked in its mountainous terrain but for the foreseeable period it will continue to suffer from shortages. Under Edward Shevardnadze, however, the republic has taken a less uncompromising attitude towards the CIS and is seeking a more balanced economic structure. With its Christian history and culture, Georgia may find it more congenial to look towards Europe and the CIS than towards some of its Muslim neighbours. The Georgians are famed for their entrepreneurial skills.

Armenia

This tiny republic, with its distinctive culture and religion, has been ethnically the least mixed of all

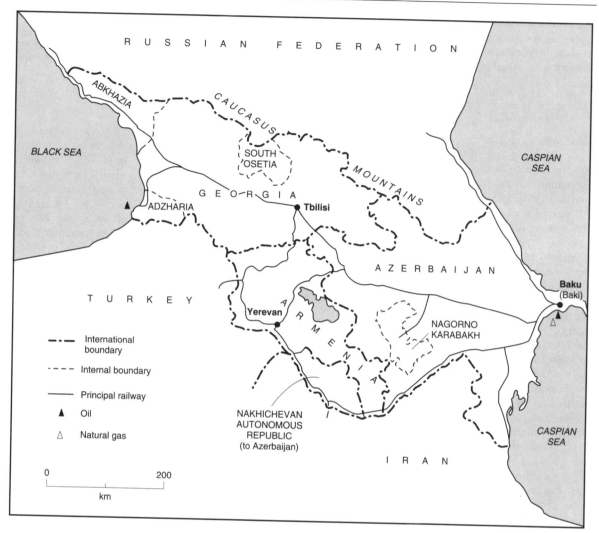

Fig. 10.4
The Transcaucasian republics.

the successor states. In 1989 Armenians constituted no less than 93 per cent of its population (88 per cent in 1959 when the Azeris formed 6 per cent). Even so it has suffered grievously from the effects of ethnic strife, notably in the long conflict with Azerbaijan over Nagorno-Karabakh. This war, together with the crises in Georgia, cut the republic off from its vital supplies of energy, raw materials and industrial equipment coming from the CIS and also from the major market for its products. The republic has had to aid the victims of the severe 1988 earthquake and the many Armenian refugees from Nagorno-Karabakh and other parts of Azerbaijan.

Armenia has various mineral resources but is very short of energy despite the availability of hydro-electric power (HEP) and a local nuclear capacity. Its economy has traditionally been oriented to machine building, light industries and specialized agriculture. The country has imported a significant proportion of its food because agriculture is restricted by the terrain. It has a good record in privatization, particularly in the agricultural sector.

Internationally, Armenia has been trying to build its relations with Russia, Turkey and Iran, the latter two being of potential significance for the republic's communications with the outside world. There is a large Armenian *diaspora* living abroad which may be significant for its future development. However, given its difficult geographical location and the continuing struggle with Azerbaijan, its immediate future looks bleak.

Azerbaijan

Turkic-speaking Azerbaijan has also suffered severely from the effects of ethnic strife and the war over Nagorno-Karabakh. In 1989, 82.7 per cent of its population was Azeri, 5.6 per cent Russian (13.6 per cent in 1959) and a similar percentage Armenian (12 per cent in 1959). Since then, however, it has received an influx of Azeri refugees from Armenia and also from Nagorno-Karabakh, and most of its Armenian population has fled. The southern part of the republic has been part of the theatre of war.

Ethnic conflict has vitiated the one great advantage which Azerbaijan enjoys by comparison with most of the states discussed so far, namely the availability of energy. Resources of oil and also gas have attracted the interest of foreign investors and have the potential to allow the republic to play an advantageous role on the world market. Much of the infrastructure, however, needs modernizing. Likewise the engineering industry, focused on the manufacture of oil industry equipment, has become run down and there is now more emphasis on the production of metal goods. Agriculture played a significant role during the Soviet period, utilizing the advantageous climatic conditions to concentrate on a varied and often specialized output. However, this was at the expense of self-sufficiency in basic food production and the sector has suffered as a result of the disappearance of the Soviet market. Partly because of its political troubles, Azerbaijan has been slow to reform its economy and the latter has been deteriorating since 1989. Agricultural land is being leased rather than privatized.

Although in a slightly more favourable geographical location than Armenia and presently Georgia, Azerbaijan does have problems communicating with the outside world and its transport infrastructure is in need of renewal. It has been dependent on Russia for the export of its oil. Relations were at first difficult because of Azerbaijan's displeasure over attitudes towards the Nagorno-Karabakh dispute. However, the Aliyev government which came to power in 1993 has been more pro-Russian and pro-CIS, and also less pro-Turkish, than its predecessor. The country has responded favourably to overtures from various Middle Eastern states like Turkey and Iran. It has become a member of the economic organization ECO, originally established by Iran, Pakistan and Turkey but now also joined by Afghanistan, Uzbekistan, Turkmenistan, Kyrgyzstan and Kazakhstan. In this way, Azerbaijan is hoping to find alternative export routes for its oil as well as making friends among its ethnically and religiously related neighbours.

Kazakhstan

Both geographically and ethnically, Kazakhstan is something of a transitional republic between Siberia and Central Asia, and thus in a sense between Europe and Asia (Fig. 10.5). Environmentally, the northern part is rather similar to neighbouring parts of West Siberia and its population is predominantly Russian. This is partly the result of a long history of Russian settlement on the steppe, and partly of more recent events like Khrushchev's Virgin Lands campaign. Indeed, before the 1989 census, Russians were the biggest ethnic group in the republic, and even in 1989 they formed 37.8 per cent of the population compared with the 39.7 per cent of the Kazakhs. The central and especially southern parts of the republic are environmentally similar to Central Asia and are mainly populated by Kazakhs, although there is a significant minority of Uzbeks in parts of the south (forming 2 per cent of the whole). Other notable minorities are the Germans, the Ukrainians and the Tatars.

Kazakhstan's ethnic diversity is a potential source of instability and means that the republic finds it difficult to secure its national identity. Kazakh nationalists have shown resentment at what they regard as a long history of Russification and at such unfortunate consequences of Soviet policy as the pollution of large territories by nuclear weapons testing. They advocate policies to secure the dominance of Kazakh linguistic and cultural norms. However, the government of President Nazarbayev has been cautious, fearful of alienating the Russian population in Kazakhstan and antagonizing chauvinistic elements in Russia itself. Thus far this policy has been very successful in avoiding the worst facets of ethnic unrest.

Kazakhstan was the USSR's second largest republic, occupying 12 per cent of the territory, and in 1988 was the fourth republic for industrial production and third for agricultural output. It is rich in energy and minerals and is thus well placed in comparison with many post-Soviet states. However, its economy was highly integrated with the rest of the USSR and it has suffered severely because of the break-up. The Nazarbayev government has been one of the most consistent advocates of CIS integration.

The republic is only one of four to inherit a nuclear capability from the USSR. Its attitude to nuclear weapons has been influenced by its proximity to China.

Fig. 10.5
Kazakhstan and the four republics of Central Asia.

Kazakhstan's transitional location and character are reflected in its foreign policy which has fluctuated between seeking closer links with the CIS and beyond it with the West, on the one hand, and with Central Asia and neighbouring parts of the Middle East, China and the rest of Asia on the other. It has pursued a policy of economic reform and marketization but sometimes with an uncertainty which seems to reflect its identity problem.

The four *Central Asian republics* are very different from much of the rest of the former USSR, being ethnically and culturally part of the Islamic and Middle Eastern world rather than part of Europe and northern Asia. As noted in previous chapters, the area came into the Russian realm only in the middle

of the nineteenth century and, since it was already well settled, Russian colonization had relatively little impact from a cultural point of view. Today this region suffers from the consequences of a rapidly expanding population against the background of aridity and environmental deterioration. The latter is largely a consequence of Soviet determination to transform it into a 'plantation economy' (Dienes 1987) supplying raw cotton and minerals to the USSR's European industrial core. Another problem derives from the way in which the Soviets subdivided the region into four republics despite the complex ethnic geography and the lack of a strong pattern of national (as opposed to regional or clan) identities. The result is that the four republics are ethnically quite intermixed (see Table 3.1) with plenty of scope for territorial disputes and centrifugal movements. Central Asia was the least developed of Soviet regions, having an agricultural and rural orientation, and faces many economic difficulties which are familiar to Third World countries. The four republics are briefly described below (see Fig. 10.5).

Uzbekistan

This is the most populous of the four republics and tends to assume the role of regional political leader. The Karimov government, which is based on the old communist hierarchy, has an authoritarian stance and has been rather intolerant of democratic and Islamic opposition groups. The Tajik minority has also suffered some discrimination as a result of political problems in neighbouring Tajikistan. A conservative approach has been taken towards economic reform. There have been attempts to diversify agriculture away from the cotton orientation but the republic can barely afford to dispense with this valuable export, notwithstanding the environmental problems and water shortages. Uzbekistan has important resources, including energy (especially gas) and non-ferrous minerals (including gold) but these are not sufficient on their own to pay for needed imports. The republic is seeking alternative markets to the CIS for its products and like its neighbours seeks foreign investment to develop rural processing activities.

Turkmenistan

Turkmenistan long had the reputation of being the least developed of the Soviet republics and an archetypal 'plantation economy'. Resentment at Soviet policy which pushed it towards cotton monoculture and apparently away from industrialization has since led to its pursuit of a rather independent stance relative to the rest of the CIS. In particular, Turkmenistan's resources of natural gas as well as oil and certain other minerals mean that its government can look to foreign trade to earn important revenues for economic development. Its environmental and general development problems are severe, however. Thus far Turkmenistan has been one of the most stable republics, partly because of its traditional social structure and partly because of its ethnic character, with generally good relations between Turkmen and Russians. The republic has been cultivating its relations with Iran and other neighbours. The authoritarian government of President Niyazov has been very intolerant of dissent and the president has encouraged a marked personality cult. His approach to economic reform is conservative.

Kyrgyzstan

Kyrgyzstan shares many of the problems of its Kazakh and Central Asian neighbours. The government of President Akayev, who is not from a communist background, has favoured a pro-Western stance and has attempted to introduce policies of economic reform and partial democratization. The presence of a significant Russian minority in the cities with a somewhat pro-Western native intelligentsia, plus the possibility of exploiting the varied mineral and hydro-electric wealth of this mountain republic, seemed to encourage this approach. Nevertheless, the country suffers from many of the characteristic problems of Central Asia and the economy has suffered in consequence. The ethnic mix and tortuous boundaries have given rise to numerous disputes. There are traditional difficulties of overdependence on cotton and rural poverty, especially in the south. Moreover, Kyrgyzstan was formerly very dependent on the rest of the USSR for imports and has suffered severely from the break-up. Despite the wish of some Kyrgyz to see their republic as an outlier of Europe, a closer relationship with some of its Asian neighbours seems more likely in the short term.

Tajikistan

Of the four Central Asian republics, Tajikistan has suffered most severely from the political and ethnic difficulties which have attended the break-up of the USSR. During 1992, these problems precipitated open civil war between pro-communist groups on the one hand and Islamic nationalist and democratic elements on the other, and there have been several changes of government. The instability is encouraged by regionalist tendencies and also by territorial disputes. The presence of a significant Uzbek minority in parts of the north and the south, forming nearly 24 per cent of the population in 1989 (though many have left), is an obvious case in point. This complicates the republic's unity and its relations with Uzbekistan. Another is the Pamiri population of the Gorno-Badakhshan territory to the east who demand autonomy for their region.

Natural resources include non-ferrous minerals and HEP, but development is hindered by lack of investment and the political instability. Tajikistan lacks energy resources apart from HEP. The republic is a classic Central Asian state suffering from increasing population pressure on the land, mounting unemployment and an overdependence on cotton and other forms of agriculture. The economy has suffered as a result of the loss of subsidies with the Soviet break-up. Economic reform began in 1992 but has since been hindered by civil war and political factors. The country is in great need of investment for rural development. Since the Tajiks are an Iranian people, Iran might be a source of help, although there are both political and religious differences between the two states.

The Russian Federation

Russia dwarfs all the other republics (Fig. 10.6). At the end of the Soviet era, it occupied 76 per cent of Soviet territory and contained 51 per cent of the population. In 1985 it accounted for 61 per cent of the industrial output of the USSR and 46 per cent of its agricultural production. Its resource endowment was formidable. At the time of maximum energy production around 1988, Russia produced about 91 per cent of the USSR's oil, 64 per cent of its gas and 55 per cent of its coal. Analogous figures for 1992 were 88 per cent, 82 per cent and 56 per cent respectively. Most other resources are also in adequate supply, with the important exception of agricultural resources. The resource wealth, however, cannot hide the fact that the Russian Federation mirrors most of the problems of the former Soviet Union as a whole, particularly its need for economic restructuring.

With almost 82 per cent of its population ethnic Russians, the Federation would seem to be in a better position as regards its ethnic integrity than many of the other post-Soviet states. However, as Chapter 3 indicated, this is hardly the case. The 1989 census recognized some 67 nationalities in Russia or even more if the northern minority peoples are distinguished as separate ethnicities. This great variety of peoples differ enormously in culture, aspiration and outlook and only long association with Russia suggests any basis for unity. The federal nature of the Russian state was examined in Chapter 3 together with some of the difficulties of maintaining a federal union. Russia's particular difficulties become apparent when it is realized that the 'autonomous territories' (the 21 republics, the Jewish Autonomous Oblast, and the autonomous okrugs) contain 17.6 per cent of the Federation's population but no less than 53.3 per cent of its territory. Thus, while attempts at secession by some of the autonomies in the North Caucasus (where Russians are generally in a minority) might not matter much, similar attempts by the strategically located ones in the Volga–Urals region like Tatarstan or Bashkortostan, or in the resource-rich north, would have much more serious implications. Russia's integrity is also threatened by the demands being made by non-ethnic units in Siberia and elsewhere for more control over their resources and their future development (Bradshaw 1992).

Uncertainty about the future of Russia relates to an apparent lack of legitimacy in the Russian state itself (Shaw 1993). The sudden break-up of the Soviet Union has left many people wondering whether Russia will not now fragment and what should be the future political relationships between the different parts of the Federation, and between the latter and the other post-Soviet republics. Many Russian nationalists regard the present frontiers of the Federation as the product of historical accident and therefore as having no inherent justification. Many lament the passing of the Soviet Union's Empire in eastern Europe and the break-up of the USSR itself. In particular, nationalists have found the independence of Ukraine and Belarus' hard to accept in view of their ethnic, cultural and historical links with Russia. There is thus a strong constituency for the reassertion of Russian power particularly within the boundaries of the former USSR. This is linked to concern, whether genuine or not, about the fate of Russians who live in the non-Russian republics. There are also widely felt worries about security. As argued in Chapter 1, Russians have worried about their security throughout history, often enough with good reason. The Soviet Union's response was to set up a series of buffer states in eastern Europe to stand between itself and the West. Not only has that buffer now gone but the danger to nationalists is that some of the post-Soviet states might also become a security risk. It is exactly the same kind of argument which many Americans have voiced about protecting 'America's backyard' from subversion. President Yeltsin has requested international recognition of Russia's right to intervene against political instability within the frontiers of the former USSR. The desire of former Soviet satellites in eastern and central Europe to join NATO has also attracted Russian suspicion. Thus, just as other post-imperial countries have found adjusting to their new, reduced role in the world far from easy, the same is now true of Russia.

At a deeper level, Russian policy towards its post-Soviet neighbours and towards the outside world raises numerous questions about how Russians now see themselves and what kind of society they aspire to be. The questions often go far back into Russian history. For many Russians, friendly and cooperative relations with the West symbolize that desire for Westernization and Europeanization which found forceful expression in the reign of Peter the Great. Advocates of marketization and democratization on the Western model can be regarded as the present-day heirs of the Westernizers of the past. Even among the marketizers, however, there are those who would see Russia's future as lying to the east in the Pacific Basin rather than in Europe. Equally, there are those who would borrow

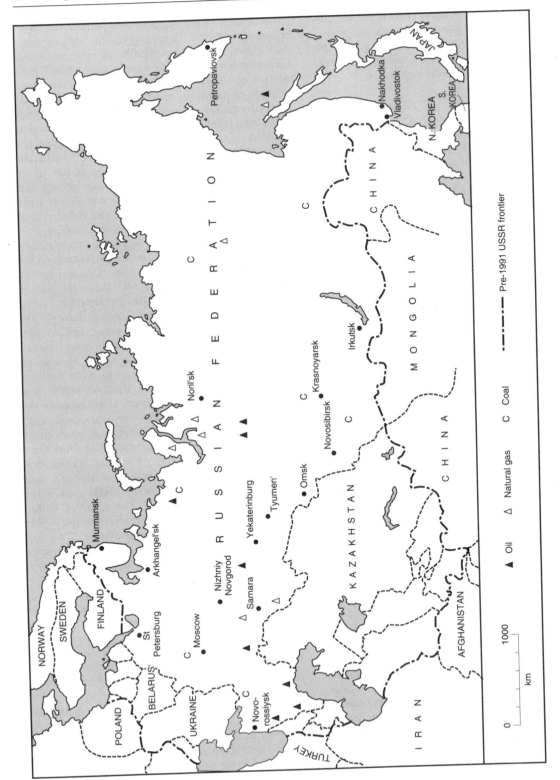

Fig. 10.6
The Russian Federation and its neighbours.

more selectively from abroad, and others who would view most if not all such borrowings with suspicion and who believe that Russia must forge its own, unique way in the world. One version of the latter position which has a long historical pedigree is the view which rejects Westernization entirely as alien to Russian tradition or as hostile to the country's true interests. Some people, for example, have argued that Russia's long associations with and geographical proximity to the Eastern Orthodox and Muslim worlds forces it to have different interests in those parts of the globe from those espoused by the West, particularly by the United States. This is an argument which appeals to many on the political right, while its rejection of the West also finds favour with communists. For Russians, then, one of the greatest challenges of the post-Soviet transition is to decide exactly what Russia is.

Future prospects

The political instability which has accompanied the current transition from communism makes it impossible to predict the future for any of the successor states. In all of them political struggles are now underway with repercussions which could radically affect their course of development. Each of the states faces choices regarding its future, but the choices finally made will depend upon the group or groups which succeed in the struggle for political power. For some states the options seem more straightforward than for others. The Baltic states in particular seem likely to continue to seek a closer relationship with northern, central and western Europe and to reduce their links with the East. Even they, however, may find it in their long-term interests to maintain some eastward connections. For other republics, and for Russia in particular, the choices seem more open: whether to try to pursue radical Westernization or to temper this with policies designed to address their unique problems and interconnections with eastern Europe, the Middle East and Asia. The Central Asian republics, with their special cultural and economic circumstances, may well find the radical Westernization option even more difficult to pursue than some of their post-Soviet neighbours. However, there is still a variety of models for consideration: the secular modernization of Turkey; the Islamic fundamentalism of Iran; the marketized socialism of China; or the authoritarian capitalism of some of the newly industrializing Asian states.

The centrifugal forces impinging upon the post-Soviet republics threaten to send them in different directions and even to pull some of them to pieces. In emphasizing those forces, however, we must not forget the forces pushing them in the opposite direction. The Commonwealth of Independent States, which was established in December 1991 to replace the USSR, now consists of the twelve post-Soviet republics apart from the Baltic states. It has attempted to address issues which appear to demand cooperation rather than confrontation to resolve. The major one is the issue of economic interrelations which has been discussed throughout this book and particularly in Chapter 9. Another is security: whether the successor states have common defence interests and whether, by cooperating, they can avoid the conflicts which might otherwise ensue. Yet another is the environmental problem bequeathed by the USSR. Pressing issues such as air pollution, the pollution of rivers and inland seas like the Baltic, Black and Caspian Seas, nuclear power and nuclear pollution, the Aral Sea crisis and the water crisis in Central Asia, are all problems which demand an international response. Thus far the CIS has been a rather weak entity with each of the successor states determined to maintain its own interests and independence, especially in its relations with Moscow. But common concerns may eventually modify this stance.

The fifteen successor states, developed together for so long, are now embarking on a new phase in their evolution. How far that evolution will still keep them together, or force them apart, remains to be seen. What can be said is that the future is likely to be a particularly difficult one for those republics which fail to grapple decisively with the problems inherited from the years of Soviet development.

References

Bradshaw M J 1992 Siberia poses a challenge to Russian federalism. *RFE/RL Research Report* **1**(41), 16 October: 6–14.

Dienes L 1987 *Soviet Asia: Economic development and national policy choices.* Westview Press, Boulder

Marples D R 1993 The post-Soviet nuclear power program. *Post-Soviet Geography* **XXXIV**(3): 172–84

Sagers M 1991 Regional aspects of the Soviet economy. *PlanEcon Report* **VII**(1–2), 15 January

Shaw D J B 1993 Geographic and historical observations on the future of a federal Russia. *Post-Soviet Geography* **XXXIV**(8): 530–40.

Shaw D J B, Bradshaw M J 1992 Problems of Ukrainian independence. *Post-Soviet Geography* **XXXIII**(1): 10–20

Stewart S 1993 Ukraine's policy towards its ethnic minorities. *RFE/RL Research Report* **2**(36), 10 September: 55–62

Glossary

ASSR: Autonomous Soviet Socialist Republic

Autonomous oblast: autonomous region

Autonomous okrug: autonomous district

BAM: Baykal–Amur Mainline Railway

CIS: Commonwealth of Independent States

CMEA: Council for Mutual Economic Assistance (also: COMECON); the economic organization which facilitated trade between the USSR and its communist allies

Dacha: a second home or cottage in the countryside

FEZ: Free Economic Zone

Five-year plan: the basic economic planning instrument during the Soviet period

Glasnost': openness, a phrase much used during the Gorbachev years (1985–91) to denote the democratization drive

Gosplan: the State Planning Committee

GULAG: the General Administration of Camps, in charge of forced labour camps during the Stalin years

IMF: International Monetary Fund

KGB: Committee for State Security (the Soviet secret police)

Kolkhoz: collective farm

Kray: territory

NEP: New Economic Policy period (c. 1921–28)

NMP: net material product

Oblast: region

Okrug: district

Perestroyka: lit. reconstruction, the phrase used to describe the reforms of the Gorbachev years (1985–91)

Rayon: district

RSFSR: Russian Soviet Federal Socialist Republic, the Russian Federation's official title during the Soviet period

Soviet: a legislative council or parliament, at national, regional or local level. After 1917, soviets provided the official framework for communist government, hence the origin of the term USSR or Soviet Union.

Sovkhoz: state farm

TPC: territorial production complex

Further reading

Owing to the extremely rapid changes occurring in the former Soviet republics, students are advised to read the daily press and current affairs journals and consult the other media as much as possible. Many news-sheets on the successor states are now available, but students will find the following particularly useful:

Current Digest of the Post-Soviet Press. Weekly. Translated press and current affairs articles. Quarterly index.

Post-Soviet Geography (formerly *Soviet Geography*). Ten issues per year. Articles and news notes, mainly by Western geographers.

Radio Free Europe/Radio Liberty Research Report. Weekly. Current affairs articles by Western analysts.

General geography

Bater J H 1989 *The Soviet Scene: A geographical perspective*. Edward Arnold, London

Bradshaw M J (ed.) 1991 *The Soviet Union: A new regional geography?* Belhaven, London

Demko G J and Fuchs R J (eds) 1984 *Geographical Studies on the Soviet Union*. The University of Chicago, Department of Geography, Research Paper no. 211

Holzner L and Knapp J (eds) 1987 *Soviet Geography Studies in our Time*. University of Wisconsin, Milwaukee

Lydolph P E 1990 *Geography of the USSR*. Misty Valley, Elkhart Lake, Wisconsin

Smith G E 1989 *Planned Development in the Socialist World*. Cambridge University Press, Cambridge

Symons L (ed.) 1990 *The Soviet Union: A systematic geography*. Second Edition. Routledge, London

Physical geography and resources

Barr B and Braden K 1988 *The Disappearing Russian Forest*. Rowman and Littlefield, Totowa, New Jersey

Gustafson G 1989 *Crisis Amid Plenty: The politics of Soviet energy under Brezhnev and Gorbachev*. Princeton University Press, Princeton

Jensen R G *et al.* 1983 *Soviet Natural Resources in the World Economy*. University of Chicago Press, Chicago.

Knystautas A 1987 *The Natural History of the USSR*. McGraw-Hill, New York

Lydolph P E 1977 *Climates of the Soviet Union*. Elsevier, Amsterdam

Conservation of resources

Feshbach M and Friendly A 1992 *Ecocide in the USSR: Health and nature under siege*. Aurum, London

Massey-Stewart J (ed.) 1992 *The Soviet Environment: Problems, policies and politics*. Cambridge University Press, Cambridge

Panel on the state of the Soviet environment at the start of the nineties 1990. *Soviet Geography* 31: 401–68

Peterson D J 1993 *Troubled Lands: The legacy of Soviet environmental destruction*. Westview Press, Boulder, Colorado

Pryde P R 1991 *Environmental Management in the Soviet Union*. Cambridge University Press, Cambridge

Ziegler C E 1987 *Environmental Policy in the USSR*. University of Massachusetts Press, Amherst

History and historical geography

Bater J H and French R A (eds) 1983 *Studies in Russian Historical Geography*. 2 vols. Academic Press, London

Blackwell W L (ed.) 1974 *Russian Economic Development from Peter the Great to Stalin*. New Viewpoints, New York

Dukes P 1990 *A History of Russia: Medieval, modern, contemporary*. Second Edition. Macmillan, London

McCauley M 1981 *The Soviet Union since 1917*. Longman, Harlow

Nove A 1972 *An Economic History of the USSR.* Penguin, Harmondsworth

Pallot J and Shaw D J B 1990 *Landscape and Settlement in Romanov Russia, 1613–1917.* Clarendon Press, Oxford

Riasanovsky N V 1977 *A History of Russia.* Third Edition. Oxford University Press, Oxford

Sumner B H 1944 *Survey of Russian History.* Methuen, London

Economy and economic geography

Aslund A 1991 *Gorbachev's Struggle for Economic Reform.* Second Edition. Pinter, London.

Ellman M and Kontorovich V (eds) 1992 *The Disintegration of the Soviet Economic System.* Routledge, London

Gregory P R and Stuart R C 1993 *Soviet and Post-Soviet Economic Structure and Performance.* Fifth Edition. Harper Collins, New York

Sagers M J and Shabad T 1990 *The Chemical Industry in the USSR: An economic geography.* Westview Press, Boulder, Colorado

Shabad T 1969 *Basic Industrial Resources of the USSR.* Columbia University Press, New York

Zumbrunnen C and Osleeb G 1985 *The Soviet Iron and Steel Industry.* Rowman and Allenhold, Totowa, New Jersey

Transport and transport geography

Ambler J, Shaw D J B, Symons L (eds) 1985 *Soviet and East European Transport Problems.* Croom Helm, London

Kontorovich V 1992 The railroads. In Ellman M, Kontorovich V (eds) *The Disintegration of the Soviet Economic System.* Routledge, London, pp. 174–92

North R N 1978 *Transport in Western Siberia: Tsarist and Soviet Development.* University of British Columbia Press, Vancouver

North R N 1991 Perestroyka and the Soviet transportation system. In Bradshaw M J (ed.) *The Soviet Union: A new regional geography.* Belhaven, London, pp. 143–64

Symons L, White C (eds) 1975 *Russian Transport: An historical and geographical survey.* Bell, London

Agriculture and rural geography

Hedlund S 1984 *Crisis in Soviet Agriculture.* Croom Helm, London

Hedlund S 1989 *Private Agriculture in the Soviet Union.* Routledge, London

Medvedev Z 1987 *Soviet agriculture.* W W Norton, New York

Stuart R C (ed.) 1983 *The Soviet Rural Economy.* Totowa, New Jersey

Wadekin K E 1989 *Communist Agriculture: Farming in the Soviet Union and eastern Europe.* Routledge, London

Planning and urban geography

Andrusz G 1984 *Housing and Urban Development in the USSR.* Macmillan, London

Bater J H 1980 *The Soviet City.* Edward Arnold, London

Morton H W and Stuart R C (eds) 1984 *The Contemporary Soviet City.* Macmillan, Basingstoke

Pallot J and Shaw D J B 1981 *Planning in the Soviet Union.* Croom Helm, London

Nationalism and nationalities issues

Denber R (ed.) 1992 *The Soviet Nationality Reader.* Westview Press, Boulder, Colorado

McAuley A (ed.) 1991 *Soviet Federalism: Nationalism and economic decentralisation.* Leicester University Press, London

Simon G 1991 *Nationalism and Policy towards the Nationalities in the Soviet Union.* Westview Press, Boulder, Colorado

Smith G (ed.) 1990 *The Nationalities Question in the Soviet Union.* Longman, London

Political, social and demographic geography

Chinn J 1977 *Manipulating Soviet Population Resources.* Macmillan, London

Desfosses H (ed.) 1981 *Soviet Population Policy: Conflicts and constraints.* Pergamon, New York

French R A 1987 Changing spatial patterns in Soviet cities – planning or pragmatism? *Urban Geography* 8(4): 309–20

Geography of human resources in the post-Soviet realm: a panel 1993. *Post-Soviet Geography* 34(4)

Herlemann H 1987 *Quality of Life in the Soviet Union.* Westview, Boulder

Lydolph P E 1989 Recent population characteristics and growth in the USSR. *Soviet Geography* 30(10): 711–29

McAuley A 1987 *Economic Welfare in the Soviet Union: Poverty, living standards and inequality.* University of Wisconsin Press, Madison

Matthews M 1986 *Poverty in the Soviet Union.* Cambridge University Press, Cambridge

Mitchneck B A 1991 Geographical and economic determinants of interregional migration in the USSR. *Soviet Geography* **32**(3): 168–89

Panel on patterns of disintegration in the former USSR 1992. *Post-Soviet Geography* **33**(6)

Smith D M 1987 *Geography, Inequality and Society.* Cambridge University Press, Cambridge

Yanowitch M 1977 *Social and Economic Inequality in the Soviet Union.* Robertson, London

Regional geography

Aves J 1993 *Post-Soviet Transcaucasia.* RIIA, London

Dienes L 1987 *Soviet Asia: Economic development and national policy choices.* Westview Press, Boulder, Colorado

Lewis R A (ed.) 1992 *Geographic Perspectives on Soviet Central Asia.* Routledge, London

Marples D R 1991 *Ukraine under Perestroika: Ecology, economics and the workers' revolt.* Macmillan, London

Rodgers A (ed.) 1991 *The Soviet Far East: Geographical perspectives on development.* Routledge, London

Schiffer J 1989 *Soviet Regional Economic Policy: The east–west debate over Pacific Siberian development.* Macmillan, London

Smith G E (ed.) 1994 *The Baltic States.* Cambridge University Press, Cambridge

Wood A (ed.) 1987 *Siberia: Problems and prospects for regional development.* Croom Helm, London

Wood A and French R A (eds) 1989 *The Development of Siberia: People and resources.* Macmillan, London

Index